Philosophy of Science in the Twentieth Century

D1082125

To my parents

Philosophy of Science in the Twentieth Century

Four Central Themes

Donald Gillies

BLACKWELL
Oxford UK & Cambridge USA

Copyright © Donald Gillies 1993

The right of Donald Gillies to be identified as author of this work has been
asserted in accordance with the Copyright, Designs and Patents Act 1988.

First published 1993

Blackwell Publishers
108 Cowley Road
Oxford OX4 1JF
UK

238 Main Street, Suite 501
Cambridge, Massachusetts 02142
USA

British Library Cataloguing in Publication Data
A CIP catalogue record for this book is available from the British Library.

Library of Congress Cataloging-in-Publication Data
Gillies, Donald.
    Philosophy of Science in the twentieth century: an introduction /
Donald Gillies.
        p.   cm.
    Includes bibliographical references and index.
    ISBN 0-631-15864-2. – ISBN 0-631-18358-2 (pbk.)
    1. Science – Philosophy – History – 20th century.   I. Title.
Q174.8.G55   1993
501 – dc20   92-36318
                                                                              CIP

Typeset in 10½ on 12pt Bembo by Best-set Typesetter Ltd., Hong Kong
Printed in Great Britain by T.J. Press Ltd, Padstow, Cornwall

This book is printed on acid-free paper

In the research work of all branches of empirical science this *spirit of a scientific conception of the world* is alive. However only a very few leading thinkers give it systematic thought or advocate its principles.

Neurath *et al.*, *The Scientific Conception of the World: The Vienna Circle*, 1929.

A certain type of penicillium produces in culture a powerful antibacterial substance. . . . The active agent is readily filterable and the name 'penicillin' has been given to filtrates of broth cultures of the mould. . . . Penicillin is non-toxic to animals in enormous doses and is non-irritant. . . . It is suggested that it may be an efficient antiseptic for application to, or injection into, areas infected with penicillin-sensitive microbes.

Fleming, *British Journal of Experimental Pathology*, 1929.

Contents

Preface   xi
Acknowledgements   xiii

Part I   Inductivism and its Critics   1

1   Some Historical Background: Inductivism, Russell and
    the Cambridge School, the Vienna Circle and Popper   3
    *1.1 Inductivism 3   1.2 The Uniformity of Nature and the
    Principle of Induction 8   1.3 Russell and the Cambridge
    School 11   1.4 The Vienna Circle 17   1.5 The Twentieth-
    Century Revolution in Physics 20   1.6 Popper 21
    1.7 The Dispersal of the Vienna Circle 23*

2   Popper's Critique of Inductivism. His Theory of Conjectures
    and Refutations (or Falsificationism)   26
    *2.1 Popper's Critique of Inductivism 26   2.2 Popper's
    Theory of Conjectures and Refutations (or Falsificationism) 29
    2.3 The Distinction between Discovery and Justification 30
    2.4 Some General Observations on Popper's Theory of
    Scientific Method 32   2.5 Kepler's Discovery of the Elliptic
    Orbits of the Planets 36   2.6 Fleming's Discovery of
    Penicillin: Creative Induction 39   2.7 The Discovery of the
    Sulphonamide Drugs: Mechanical or Baconian Induction 48*

3   Duhem's Critique of Inductivism   54
    *3.1 Inductivism as the Newtonian Method 54   3.2 Newton's
    Inference of the Law of Gravity from Kepler's Laws and
    Duhem's Objections 58   3.3 Criticisms of Inductivism and the
    Revolution in Physics 60   3.4 The Lives of Duhem and
    Poincaré 63   3.5 Artificial Intelligence and the Revival of
    Inductivism 69*

Part II   Conventionalism and the Duhem–Quine Thesis   73

4   Poincaré's Conventionalism of 1902   75
     4.1 Kant's Philosophy of Geometry 75   4.2 The Discovery
     of Non-Euclidean Geometry 77   4.3 Poincaré's
     Conventionalist Philosophy of Geometry 85   4.4 Poincaré's
     Conventionalism and Newtonian Mechanics 90
     4.5 Poincaré on the Limits of Conventionalism 94

5   The Duhem Thesis and the Quine Thesis   98
     5.1 Preliminary Exposition of the Thesis. The Impossibility of
     a Crucial Experiment 98   5.2 Duhem's Criticisms
     *of Conventionalism. His Theory of Good Sense (*le bon
     sens*) 102   5.3 The Quine Thesis 108   5.4 The*
     Duhem–Quine Thesis 112

Part III   The Nature of Observation   117

6   Protocol Sentences   119
     6.1 Carnap's Views on Observation Statements in the Early
     1930s 120   6.2 Neurath's Views on Observation Statements
     in the Early 1930s 122   6.3 Popper's Views on Basic
     Statements in 1934 124

7   Is Observation Theory-Laden?   132
     7.1 Duhem's View that All Observation in Physics is Theory-
     Laden 132   7.2 A Reinforcement of the Holistic Thesis
     and Neurath's Principle 137   7.3 Some Psychological
     Findings 140   7.4 Some General Conclusions 146

Part IV   The Demarcation between Science and Metaphysics   151

8   Is Metaphysics Meaningless? Wittgenstein, the Vienna
     Circle, and Popper's Critique   153
     8.1 Introduction: The Demarcation Problem and its Importance
     153   8.2 Wittgenstein's Life 157   8.3 Wittgenstein's
     Tractatus 165   8.4 The Vienna Circle on Metaphysics 172
     8.5 Popper's Critique of the Vienna Circle on Metaphysics 177
     8.6 Wittgenstein's Later Theory of Meaning 181   8.7 The
     Influence of Wittgenstein's Life on his Philosophy 185

9   Metaphysics in Relation to Science: The Views of
     Popper, Duhem, and Quine   189
     9.1 Popper on Metaphysics in Relation to Science 189
     9.2 Duhem and Quine on the Status of Metaphysics 192

9.3 Duhem and Popper on the Influence of Metaphysics on Science 195   9.4 Duhem's Defence of Religion 201

10 Falsificationism in the Light of the Duhem–Quine
    Thesis   205
    *10.1 Falsificationism and the Falsifiability Criterion 205
    10.2 Existential Statements 206   10.3 Probability
    Statements 207   10.4 Falsifiability and the Duhem–Quine
    Thesis 210   10.5 A Suggested Demarcation Criterion
    involving the Principle of Explanatory Surplus 214
    10.6 How Much of Falsificationism can be Retained? 221
    10.7 Some Concluding Philosophical Remarks 228*

Notes   231
References   238
Index   245

Preface

This book gives a history of philosophy of science in the twentieth century. Naturally, in so short a space, it is not possible to cover everything, and I have accordingly selected what seem to me to be the most central themes. Inevitably there is a personal element in this choice. Other writers on the subject might perhaps have selected a slightly different list of topics; but I think all would agree that the four central themes discussed here are indeed of the greatest importance.

The book, I hope, contains some things that will be of interest to those already familiar with the subject, but it is also intended as an introduction to philosophy of science. Indeed, the historical approach seems to me an excellent way of introducing almost any subject, since a good understanding of current ideas can be obtained through the study of the developments which led up to them. Since the book is intended as an introduction, I have not presupposed any previous knowledge of philosophy or of science, and have avoided, wherever possible, mathematical formulas and other such technicalities. This is not to say that the philosophical arguments are always easy to follow, but at least I have tried to avoid creating any artificial barriers to understanding which might be raised by more than the minimal use of formal logic or mathematical symbolism.

The nature of science and the philosophical problems to which it gives rise constitute the subject-matter of the philosophy of science. Ideas in the philosophy of science thus become empty unless they can be related to scientific practice. In the course of the book, therefore, I have described a number of episodes from either recent or more historical science, in order to illustrate the philosophical discussions. These episodes include Kepler's discovery of the elliptic orbits of the planets, the development of non-Euclidean geometry, Fleming's discovery of penicillin, and the introduction of quantum theory by Planck and Einstein. Although these examples are introduced to make specific philosophical points, they have, I believe, an interest of their own. Most intellectuals are more familiar with the works and lives of great artists than with those of great scientists. It is understandable that this should be so; yet the achievements of a Kepler, a Fleming, or an Einstein are as remarkable as those of a Mozart, a Proust, or a Michelangelo.

I have included in the book not just an account of the ideas of the philosophers, but some details about their lives as well. These were often surprisingly dramatic and interesting. Philosophy of science is not such a remote and 'ivory tower' pursuit as it might at first seem. Questions about science impinge on both religion and politics, and are therefore liable to arouse all the passions which these activities inspire. This is clearly seen in the fate of the Vienna Circle, perhaps the most influential group of philosophers of science in the twentieth century. The Circle in fact met in Vienna only for about twelve years, from 1922 to 1934. In 1934 its leader, Professor Moritz Schlick, was assassinated by a Nazi student, and the other members of the Circle were forced to flee the city.

The book has taken several years to write, and I have benefited from the help of numerous friends and colleagues. So as not to overburden the preface, I try to express my thanks in a special Acknowledgements section which follows.

Donald Gillies,
Department of Philosophy,
King's College London,
Strand,
London WC2R 2LS
June 1992

Acknowledgements

Anyone who writes a general book like this one needs a great deal of help from specialists in different areas, and in this respect I have been extremely fortunate.

My discussion of the twentieth-century revolution in physics, and more specifically of the science and philosophy of science of Henri Poincaré has benefited from many long discussions with Jerzy Giedymin and Arthur Miller. It should be added that these two distinguished scholars have quite opposite views on many points. I have tried to give some indications of these differences, as well as references to the relevant papers and books, so that an interested reader can learn more about this fascinating controversy.

In my treatment of Duhem, I have been assisted by two distinguished Duhem scholars, Niall Martin and Anastasios Brenner. Niall Martin I have known for very many years, and large indeed must be the number of conversations we have had about Duhem. But I have also been able to use his recently published book. I became acquainted with Anastasios Brenner only through studying his recent paper and book on Duhem, but he was kind enough to read some of the sections of this book dealing with Duhem and to offer very helpful comments.

My original training was in the mathematical sciences. However, a former colleague, Harmke Kamminga, not only read through some early drafts of the book and offered useful comments, but insisted, in this and other contexts, that an adequate philosophy of science can be achieved only by taking account of the biological as well as the physical sciences. It was her exhortation which made me decide to include some examples from biochemistry. Another colleague, Melvin Earles, led me to the examples of the discovery of penicillin and the sulphonamide drugs. He gave me every assistance in my task by lending me books, reading through earlier drafts and correcting mistakes, and drawing my attention to important papers. I was also very fortunate to meet the late Peter Mitchell, Nobel prize-winner for chemistry in 1978, when he came to talk to our seminar. Peter Mitchell's interest in the philosophy of science and great kindness of heart led him to take the trouble to explain some details of his chemiosmotic hypothesis to a novice like myself and to read and comment on my drug discovery examples. He was also kind enough to invite me to the meeting in October 1990 for the silver jubilee of the laboratory at Glynn where he carried out the researches which led to the Nobel prize. On this occasion I met Joseph Robinson and Bruce Weber, both biochemists with a strong interest in history and philosophy of science, with whom I have had much useful discussion and correspondence. It was with great sorrow that I learned of the sudden death of Peter Mitchell in April 1992.

As regards the results from empirical psychology mentioned in my discussion of the nature of observation, I was greatly aided by Richard Gregory who made many useful suggestions and provided me with the photograph used for Plate 4.

The general structure of the book emerged as the result of a series of long and most enjoyable conversations with Giulio Giorello, who also made many helpful suggestions on specific points. Nearly all the material has been presented as lectures, both to philosophy students in London University's intercollegiate lecture programme and as part of philosophy of science courses for science students at King's College London. I have benefited from many astute comments from these lively and critical audiences.

My own views on history and philosophy of science have been much influenced over the years by the weekly seminars in the subject first at Chelsea College and then, after its merger, at King's College London. We have had a wonderful array of visiting speakers, including some of the greatest names in both science and history and philosophy of science. Yet the discussions, both in the seminar itself and over coffee afterwards, have often been as illuminating as the papers themselves, and I would therefore like to say a special word of thanks to the seminar's 'regulars' who may recognize the influence of some of their own ideas and comments in the pages which follow.

Finally, I should like to express my thanks to the following copyright-holders for their kind permission to use photographs. Plate 1 is Fleming's original photograph of his penicillin culture plate, and is reproduced with the permission of St Mary's Hospital Medical School. Plate 2 comes from Max Born's *Atomic Physics*, and is reproduced with the permission of the publishers of the current (8th) edition, Dover Publications, Inc. Plate 3 comes from Richard Gregory's *The Intelligent Eye*, and is reproduced with the permission of both the author and the publishers, George Weidenfeld and Nicolson. Plate 4 is from a photograph of Richard Gregory's, which he kindly sent me and gave me permission to use.

Part I
Inductivism and its Critics

1

Some Historical Background: Inductivism, Russell and the Cambridge School, the Vienna Circle and Popper

1.1 Inductivism

The first theme to be dealt with is *Inductivism and its Critics. Inductivism* is a theory of scientific method which was developed long before the twentieth century. Its main features were introduced in the seventeenth and eighteenth centuries, and it was perhaps the dominant account of scientific method in the nineteenth century. Inductivism has been widely held and developed during the twentieth century as well. Yet criticism, rather than acceptance, of inductivism has perhaps been more characteristic of twentieth-century thought. The two main critics of inductivism have been Popper and Duhem. Popper's arguments on this subject will be discussed in chapter 2, and Duhem's in chapter 3. In the present chapter, however, we will deal with the historical origins of inductivism and with some of its supporters in the twentieth century: namely, Russell and the Cambridge school and the Vienna Circle.

Inductivism as a theory of scientific method goes back to Francis Bacon (1561–1626). Bacon was a contemporary of Shakespeare (1564–1616), and there is even a palpably absurd theory that Bacon wrote Shakespeare's literary works as well as his own philosophical ones. Bacon came from a family with connections among the great and powerful of the land, and he entered Parliament at the age of twenty-three intending to follow a political career. He did not meet with much success at first, for he never seems to have gained the favour of Queen Elizabeth, despite her taste for intelligent men. His fortunes improved under King James, despite that monarch's predilection for handsome, but unintelligent, men. Bacon became Keeper of the Great Seal in 1617 and Lord Chancellor in 1618. Only two years later, however, he was prosecuted for taking bribes. Bacon did not deny the charge, but claimed that the bribes never influenced his decisions – a somewhat curious defence, indicative of Bacon's philosophical turn of mind. Bacon was condemned, and forced to leave the public arena. He devoted the rest of his life to philosophy and scientific experimentation. The latter activity proved fatal, since he died of a chill caught when stuffing a chicken with snow to see if it could be thus preserved. As so often, Bacon seems here prophetic of methods which would later come to be widely used. He wrote a number of philosophical works, but we shall rely mainly on what is perhaps his most famous: the *Novum Organum*, published in 1620 and dedicated to his patron, King James. Aristotle's collected writings on logic had been given the name *Organum* (literally 'tool'); thus Bacon's *Novum Organum*, or new *Organum*, was intended to supersede Aristotle as an instrument for reasoning.

Bacon was a man who loved money and luxury; but he was not selfish, and wished the benefits of comfort and ease to be more widely extended among humanity. He saw that this could be achieved only by the improvement of technology, and thus became one of the first to argue that scientists and technologists should be praised and supported. As he says:

> The introduction of famous discoveries appears to hold by far the first place among human actions; . . . For the benefits of discoveries may extend to the whole race of man, . . . let a man only consider what a difference there is between the life of men in the most civilized province of Europe, and in the wildest and most barbarous districts of New India; . . . And this difference comes not from soil, not from climate, not from race, but from the arts. (Bacon, 1620, p. 300)

Although Bacon's aim is largely to improve technology ('the arts', as he calls it), he does not make the mistake of some modern politicians of supposing that the best way to achieve this is to fund only projects which are capable of yielding practical applications in two or three years. On the contrary, Bacon says explicitly: 'For though it be true that I am principally in pursuit of works and the active department of the sciences, yet I wait for harvest-time, and do not attempt to mow the moss or to reap the green corn' (p. 251). He believed that in the long run technology could only be improved by improving our knowledge of the natural world, by carrying out basic research in science, as we would now say. In a famous passage he puts this idea as follows: 'Human knowledge and human power meet in one; for where the cause is not known the effect cannot be produced. Nature to be commanded must be obeyed; and that which in contemplation is as the cause is in operation as the rule' (p. 259). Bacon makes the same distinction in another way by contrasting what he calls *experiments of fruit* with *experiments of light*. As he says:

> All industry in experimenting has begun with proposing to itself certain definite works to be accomplished, and has pursued them with premature and unseasonable eagerness; it has sought, I say, experiments of Fruit, not experiments of Light; not imitating the divine procedure, which in its first day's work created light only and assigned to it one entire day; on which day it produced no material work, but proceeded to that on the days following. (p. 245)

Bacon was convinced that basic research in the natural sciences, or 'experiments of Light' as he put it, would reveal all kinds of unknown phenomena which could be used as the basis for new technologies. As he says: 'There is therefore much ground for hoping that there are still laid up in the womb of nature many secrets of excellent use, having no affinity or parallelism with anything that is now known, but lying entirely out of the beat of the imagination, which have not yet been found out' (p. 292).

The last 370 years have shown how well justified was Bacon's advocacy of scientific research with the aim of improving technology, while his hope that 'there are still laid up in the womb of nature many secrets of excellent use' has been fulfilled to an extent greater than perhaps Bacon himself could have imagined.

Bacon was not content with urging that more scientific research should be carried out. He proposed a method which, if followed, would in his view result in an expansion of our knowledge of the natural world. Some of the precise details of Bacon's method – for example, his *Tables and Arrangements of Instances* – are no longer of great interest. However, his general approach still has supporters today. It is this general approach which I will call *inductivism*. I will next turn from the particular case of Bacon to a general exposition of the main features of inductivism, but it will be convenient to illustrate this from time to time with some further quotations from the *Novum Organum*.

The basic idea of inductivism is that science starts with observations, and moves on from them to generalizations (laws and theories), and predictions. Good scientists, following the 'inductive method', will begin by making a large number of careful observations. They will then cautiously infer some generalization from the data collected, and perhaps make a prediction on the basis of this generalization. The inductive method is strongly advocated by Sherlock Holmes. Thus, in 'The Scandal in Bohemia', Holmes says to Watson: 'It is a capital mistake to theorise before one has data.' Holmes's method is to collect the data, and then infer from it what happened. This is typical of the inductive approach. It is also characteristic of inductivism to condemn speculation in the absence of data. Indeed, Bacon named such a procedure *Anticipations of Nature*, and contrasted it unfavourably with his own inductive method, which he called *Interpretation of Nature*. As he says: 'The conclusions of human reason as ordinarily applied in matter of nature, I call for the sake of distinction *Anticipations of Nature* (as a thing rash or premature). That reason which is elicited from facts by a just and methodical process, I call *Interpretation of Nature*' (Bacon, 1620, p. 262). As we shall see in the next chapter, Popper, in conscious opposition to Bacon, makes 'anticipations . . . rash or premature' an integral part of scientific method.

Let us next illustrate inductivism by two examples. The first is concerned with birds, and is a real favourite among philosophers studying the nature of human knowledge. Consider the following two generalizations:

R: All ravens are black.
S: All swans are white.

R was obtained, according to the inductivists, by observing a large number of ravens and noting that they were all black. It was then inferred that all ravens are black; and similarly in the case of swans.

However, there is an interesting difference between the two cases. R, so far as we know, is true. S was thought to be true by Europeans up till the eighteenth century. But the early explorers of Australia observed black swans, and hence refuted the generalization that all swans are white. This refutation or falsification will be important for Popper, as we shall see. Inductivists interpret it as showing that scientific inferences never produce certainty. They think, none the less, that such inferences can produce a high degree of probability.

While swans and ravens provide nice simple examples of empirical generalizations, if we want to produce a realistic theory of science and scientific method, it is desirable to analyse examples of important scientific discoveries drawn from the history of science. I will begin here by describing one such example: Kepler's discovery that planets move in ellipses with the Sun at one focus.

The story is briefly this. Between 1576 and 1597, Tycho Brahe, a Danish astronomer, made a long series of careful observations of the heavens, particularly of the movements of the planets. The telescope had not then been invented, but Tycho Brahe's observations were the most accurate ever made with the naked eye. In 1597 he left Denmark to become Imperial Mathematician at the court of the Emperor Rudolf II in Prague. Here in 1600 he took on Johannes Kepler as his assistant.

Kepler set himself the task of working out the orbit of Mars from Tycho Brahe's data. At first he thought it would take him a week. In fact, he worked on the problem for over six years, before concluding that the orbit was an ellipse. He published this result in his *Astronomia Nova* ('New Astronomy') in 1609.

At first sight this appears to be a classic example of inductive method. Tycho Brahe painstakingly makes a long series of careful observations of the planet Mars. Kepler in an equally careful and painstaking fashion infers from these observations that the orbit of Mars is an ellipse. We shall see later what an anti-inductivist, such as Popper, might say about this example.

Bacon (1620, p. 298) explains the inductive method by a striking analogy with wine-making. To make wine, we first industriously gather 'countless grapes . . . ripe and fully seasoned'. From these grapes, the juice for the wine is 'then squeezed in the press'. The grapes correspond to the observations from which scientific generalizations (laws or theories) are somehow squeezed out. But how exactly are scientific theories to be inferred from observational data? This is the next question which we must consider.

The most hopeful view is that we might obtain scientific theories by logical deduction from the data. Let us suppose that George is the name of a pet raven. Then we can give the following example of a logical deduction:

All ravens are black.	(1)
George is a raven.	(2)
Therefore, George is black.	(3)

From the premisses (1) and (2), we deduce logically the conclusion (3).

The Scottish philosopher David Hume (1711–76) pointed out that scientific inferences from observations to laws or predictions cannot be obtained by logical deduction. His argument was roughly this. If we accept the premisses of a logical deduction as true, then we must accept the conclusion as true as well. If we believe that all ravens are black and that George is a raven, then we simply must believe that George is black. The conclusion of a logical deduction follows with certainty from the premisses.

Let us contrast this with the case of inferring a law or a prediction from observations. Suppose we have observed several thousand ravens and found them all to be black. We then infer either the law that all ravens are black or the prediction that the next observed raven will be black. We cannot, however, infer this prediction with certainty from the data. We might have an experience like that of the early explorers of Australia, and find that the next raven we observe is not black at all. Similarly, we cannot infer with certainty that all ravens are black. Thus scientific inference, Hume argued, cannot be the same as logical deduction. This conclusion is supported by modern logical theory.

This situation explains the introduction of the term 'induction'. Since predictions and laws cannot be obtained from data by *deduction*, or *deductive inference*, it is claimed that they are obtained by *induction*, or *inductive inference*.

1.2   The Uniformity of Nature and the Principle of Induction

Inductivism then leads to the view that there are such things as inductive inferences; but this in turn raises quite a number of difficult questions: what is the nature of these inductive inferences? how do they differ from deductive inferences? and how can they be justified? Many of these questions were taken up by the Cambridge school, which flourished in the first few decades of the twentieth century, and then by the Vienna Circle, which was influenced by the Cambridge school and continued its work.

In chapter 6, 'On Induction', of his book *The Problems of Philosophy*, published in 1912, Bertrand Russell gives a very clear account of some of the questions concerned with inductive inference which were preoccupying him and other members of the Cambridge school at that time. Russell's thinking on this subject involved two important concepts: *the uniformity of nature* and *the principle of induction*. Let us briefly examine his treatment of each of these in turn.

Regarding the uniformity of nature, he writes:

> The problem we have to discuss is whether there is any reason for believing in what is called 'the uniformity of nature'. The belief in the uniformity of nature is the belief that everything that has happened or will happen is an instance of some general law to which there are *no* exceptions. . . . The business of science is to find uniformities, such as the laws of motion and the law of gravitation, to which, so far as our experience extends, there are no exceptions. In this search science has been remarkably successful, and it may be conceded that such uniformities have held hitherto. This brings us back to the question: Have we any reason, assuming that they have always held in the past, to suppose that they will hold in the future? (1912, pp. 98–100)

It is interesting to note that in this passage Russell refers to Newton's law of gravitation. Three years later, in 1915, Einstein published his general theory of relativity, which predicted significant deviations from Newton's law of gravitation. One of these deviations was confirmed in the eclipse experiment of 1919. Russell himself took a great interest in all these developments.

Although Russell was anxious to justify belief in the uniformity of nature, he suffered from doubts which are memorably illustrated by his chicken example: 'Rather crude expectations of uniformity are liable to be misleading. The man who has fed the chicken every day throughout its life at last wrings its neck instead, showing that more refined views as to the uniformity of nature would have been useful to the chicken' (1912, p. 98).

Russell's reflections on the chicken lead him to conclude that we should seek probability rather than certainty:

> The most we can hope is that the oftener things are found together, the more probable it becomes that they will be found together another time, and that, if they have been found together often enough, the probability will amount *almost* to certainty. It can never quite reach certainty, because we know that in spite of frequent repetitions there sometimes is a failure at the last, as in the case of the chicken whose neck is wrung. Thus probability is all we ought to seek. (p. 102)

Russell now goes on to present his positive view that inductive inferences are justified by an appeal to what is called the *principle of induction*. Russell formulates this principle as follows:

The principle we are examining may be called the *principle of induction*, and its two parts may be stated as follows:

(a) When a thing of a certain sort A has been found to be associated with a thing of a certain other sort B, and has never been found dissociated from a thing of the sort B, the greater the number of cases in which A and B have been associated, the greater is the probability that they will be associated in a fresh case in which one of them is known to be present;

(b) Under the same circumstances, a sufficient number of cases of association will make the probability of a fresh association nearly a certainty, and will make it approach certainty without limit. (p. 103)

This formulation of the principle of induction deals with predictions e.g. the next observed raven will be black. Russell (1912, pp. 104–5) gives a similar formulation dealing with laws and generalizations e.g. all ravens are black. It was pointed out to me that Russell's formulation of the principle of induction contains an error. Let A be raven, and B be black. Applying Russell's principle, we reach the conclusion that it is highly probable that A and B 'will be associated in a fresh case in which one of them is known to be present'. Thus if I know that an object is black, I should apparently conclude that it is likely to be a raven, which is clearly wrong.

This unfortunate consequence can be eliminated by rephrasing the final clause of (a) above as follows: 'the greater is the probability that they will be associated in a case in which A is known to be present'. This slip of Russell's is indicative of the great difficulty of formulating the principle of induction in a satisfactory manner.

The principle of induction is none the less regarded as fundamental by most thinkers of the inductivist camp. One such thinker was Hans Reichenbach, an associate of the Vienna Circle. To illustrate Reichenbach's adherence to the principle of induction, Karl Popper (1934, pp. 28–9) gives the following two quotations from him, both written in 1930 and published in the Vienna Circle's journal *Erkenntnis*:

This principle determines the truth of scientific theories. To eliminate it from science would mean nothing less than to deprive science of the power to decide the truth or falsity of its theories. Without it, clearly, science would no longer have the right to distinguish its theories from the fanciful and arbitrary creations of the poet's mind.

And again: 'The principle of induction is unreservedly accepted by the whole of science and . . . no man can seriously doubt the principle in everyday life either.'

Despite these encomia, the principle of induction has been severely criticized by Popper, as we shall see in the next chapter. Before pursuing these philosophical arguments, however, let us pause to fill in some more historical background. Why were Russell and his circle led to take an interest in problems connected with induction? And who were the Vienna Circle who continued and developed the work of the Cambridge school?

1.3   Russell and the Cambridge school

Bertrand Russell (1872–1970) was a scion of the English Whig aristocracy. Both his parents died before he was four years old, and he was brought up by his paternal grandparents. His grandfather (Lord John Russell) had been three times Prime Minister. Russell went to Trinity College, Cambridge, in 1890, where he became a wrangler in mathematics and then turned to philosophy. He was a Fellow of Trinity from 1895 to 1901 and a lecturer in philosophy there from 1910 to 1916.

Russell began as a Kantian, and his first book, *An Essay on the Foundations of Geometry*, published in 1897, was an attempt to give an account of non-Euclidean geometry from a Kantian standpoint. After the completion of this work, Russell became a Hegelian for a brief period, but he soon abandoned this position under the influence of his friend G. E. Moore. After these flirtations with German thinkers, Russell returned to the British tradition, and for the rest of his life held to some form of empiricism, though the particular form which he espoused varied from time to time.

Russell's first training had been in mathematics, and mathematics poses a severe problem for empiricism. According to empiricism, all knowledge is based on experience. Yet mathematical truths – for example, $5 + 7 = 12$ – seem to be known independently of experience. In the nineteenth century, John Stuart Mill, who was actually Russell's lay godfather, had suggested that, contrary to first appearances, mathematical truths such as $5 + 7 = 12$ are actually inferred inductively from observations, just like other scientific generalizations. Russell did not find this solution satisfactory, and looked for another approach in the logical works of Peano and Frege (for details see Gillies, 1982).

The basic idea was to reduce mathematics to logic. Let us consider a logical truth such as 'All bachelors are unmarried'. We do not have to conduct an empirical survey of bachelors to check that this is correct. Once we understand the definition of bachelor as un-

married man, it becomes obvious that all bachelors are unmarried.
On the other hand this truth is empty – a mere linguistic convention,
or a truth-by-definition.

The claim which Russell hoped to establish was that mathematics
can be reduced to logic – a view known as *logicism*. If logicism is
correct, then mathematical truths are basically of the same kind as
the truth that all bachelors are unmarried. So we can know
mathematical truths independently of experience, but only because
such truths are essentially empty – mere linguistic conventions
rather than assertions about the world. This approach leads to a new
version of empiricism, which could be called *logical empiricism*.
Knowledge is divided into two kinds. The first kind is knowledge
of logical truths. This knowledge is indeed independent of experience, but it consists of mere truisms, empty truths-by-definition.
Mathematical knowledge is of this kind. The second kind comprises
all really significant knowledge about the world, and is based on
experience.

The first step in developing logical empiricism is to establish that
mathematics can be reduced to logic. Frege had already tried to
show that arithmetic could be reduced to logic. However, Russell
discovered that Frege's logical system contained a contradiction –
now known as Russell's paradox. Russell's paradox can be stated
informally as follows. Consider sets, such as the set of men or the
set of fingers on the left hand of the President of the United States.
These sets are not members of themselves. Thus the set of men is a
set and not a man, and so is not a member of the set of men.
Similarly, the set of fingers on the President's left hand is not a
finger, and so not a member of itself. On the other hand, some sets
are members of themselves: for example, the set of all sets. Let us
therefore consider the set of all sets which are not members of
themselves. Is *this* set a member of itself? If it is, it isn't. If it isn't, it
is. This is a contradiction.

The discovery of this paradox in 1901 inaugurated Russell's most
creative period in theoretical philosophy, which lasted till the
outbreak of the First World War. Russell's programme was to
reconstruct Frege's logicism in such a way as to avoid the contradiction. In the course of carrying out this task, he devised his theory
of descriptions (published in 1905) and his theory of types (published in 1908). The deduction of mathematics from purely logical
premises was carried out in the three huge volumes of *Principia
Mathematica* written with Alfred North Whitehead and published
between 1910 and 1913.

The appearance of *Principia Mathematica* might well have seemed at the time to establish the truth of logicism beyond reasonable doubt. In retrospect we can see that there were difficulties from the start. It was not clear that Russell's complicated theory of types was really part of logic, and Russell had to make use of some axioms – for example, the axioms of infinity and choice – which again did not seem entirely logical in character. A German mathematician, Ernst Zermelo, had proposed an alternative method for dealing with Russell's contradiction. His approach, known as axiomatized set theory, was also published in 1908. Initially, Russell's theory of types was the more powerful system, but Zermelo's approach was strengthened with a new axiom by Thoralf Skolem and Abraham Fraenkel in 1922. Thereafter axiomatic set theory proved to be simpler and more powerful, and has been widely accepted as a foundation for mathematics, though the theory of types still retains a certain importance. The worst blow to Russell's logicism came, however, with the publication in 1931 of the paper 'On Formally Undecidable Propositions of *Principia Mathematica* and Related Systems I' by the Czech mathematician Kurt Gödel, a member of the Vienna Circle. We can state Gödel's result in the somewhat simpler form produced by the American logician John Rosser in 1936. According to the Gödel–Rosser theorem, it is possible to construct a proposition of arithmetic which is not provable in *Principia Mathematica* (*PM*) provided *PM* is consistent, but which can nonetheless be shown to be true by an informal argument outside *PM*. This proposition is thus a truth of mathematics – indeed, of arithmetic – which cannot be derived from Russell's logical axioms (assuming these are consistent). The situation cannot be repaired simply by adding a few more axioms to *PM*, for Gödel's proof can be repeated for this enlarged system. Gödel's incompleteness theorem shows that no logicist system of the kind proposed by Russell can ever be adequate for the whole of mathematics.

Since Gödel, most philosophers have abandoned logicism, though there are still a few who hope to revive it in a modified form. Between 1910 and 1930, however, logicism seemed to be an account of mathematics, in which there were certainly some difficulties, but which, on the whole, was highly plausible. Thus, under the influence of Russell, the Vienna Circle espoused logicism in the 1920s. Their viewpoint is well expressed in Rudolf Carnap's defence of logicism (Carnap, 1931) in a symposium on the foundations of mathematics published in the Vienna Circle's journal *Erkenntnis* in 1931.

Let us return now to 1912, when Russell published his discussion of induction in *The Problems of Philosophy*. At this stage Russell thought that he and Whitehead had established logicism to be an adequate account of mathematics. The next step in the development of the new logical empiricism was evidently to show that scientific, as opposed to mathematical, knowledge could be based on experience. If, however, scientific knowledge is obtained from observation by inductive inferences (as inductivism suggests), this raises the problem of how such inferences can be justified.

As we have seen, Russell's response to this problem was to postulate a 'principle of induction' and to emphasize the role of probability. Russell himself did not pursue these investigations much further, since, with the outbreak of the First World War, his interest shifted from theoretical philosophy to radical politics. During the First World War he campaigned for pacifism, and, as a result, his lectureship in philosophy at Trinity College, Cambridge, was not renewed in 1916, and he was imprisoned for six months in 1918. In the remaining fifty-two years of his long life, Russell did occasionally return to theoretical philosophy, but on the whole he devoted himself to social and political questions. He was imprisoned again, this time for a week, in 1961 at the age of eighty-nine, for his part in the campaign for nuclear disarmament.

But investigations into probability and induction were carried on at Cambridge by a series of younger philosophers of science: W. E. Johnson, John Maynard Keynes (before he turned to economics),[1] Harold Jeffreys, and Frank Ramsey. These thinkers all adopted an approach to the problem known as *Bayesianism*, a theory which still has a large number of supporters today. Now many (perhaps even most) inductivists are Bayesians, while many (perhaps even most) Bayesians are inductivists. Thus the two theories are often placed together under a single heading. It seems to me, however, that they should be carefully distinguished, since it is in fact possible to be an inductivist but not a Bayesian, and vice versa. To clarify the difference between *inductivism* and *Bayesianism*, let us briefly examine the history of Bayesianism. This theory, though not as old as inductivism, was introduced long before the twentieth century. It first appeared, in fact, in the eighteenth century.

Bayesianism is named after the English mathematician Thomas Bayes (1702–61), whose important contribution to the theory of probability was published posthumously in 1763. Bayes's paper was communicated to the Royal Society by his friend Richard Price (1723–91). Price wrote an introductory letter and an appendix. As these are significant additions, the paper could with justice be considered a joint paper, and Price should perhaps be recognized, along with Bayes, as a founder of the Bayesian approach.

Price was strongly influenced by the discussions of induction in Hume, 1748. As I have already explained, Hume argued that a generalization such as 'All ravens are black' or a prediction such as 'The next observed raven will be black' cannot be obtained by logical deduction from reports on the observation of any number of black ravens, however large. Even if thousands and thousands of black ravens have been observed, it may happen that the very next raven we encounter will have some other colour.

Price thought that Bayes's probability calculations could be used to resolve these problems posed by Hume.[2] The idea is a simple one. Observational evidence can never render a prediction or a generalization *certain*, but it may be able to make either or both *probable*. Indeed, we might be able to use the mathematical theory of probability to calculate the probability which evidence gives to a prediction or a generalization – to calculate, for example, the probability that a patient has a particular disease given that he or she displays a particular set of symptoms. The Bayesian school has devised ways to carry out such calculations, and these make use of what is known as Bayes's theorem. The opponents of Bayesianism do not deny the validity of Bayes's theorem, which is a result in the standard mathematical theory of probability. What they question is the legitimacy of the use made of this theorem by Bayesians.

Having explained the basic ideas of *Bayesianism*, we can now compare it to *inductivism*. Inductivism is a theory of how scientific research should be conducted. It claims that a scientist should make a large number of careful observations, and then obtain predictions and generalizations by the process of inductive inference from these observations. Bayesianism, on the other hand, is a suggestion as to how scientific generalizations or predictions should be assessed relative to the evidence used to support them. The Bayesian urges that the mathematical theory of probability should be used to calculate the probability of the generalization or prediction given the evidence. It is therefore clearly possible to be an inductivist without being a Bayesian. Someone might believe that the inductive method is the correct way to carry out scientific research, but have no concern for trying to calculate the probabilities of any generalizations or predictions obtained. This, indeed, was Bacon's position. In the *Novum Organum*, Bacon gives his classic account of inductivism; yet nowhere does he suggest methods for calculating the probabilities of generalizations or predictions, nor does he even suggest that this should be done. As a matter of fact, it would have been historically impossible for Bacon to have been a Bayesian. Bayesianism is all about a possible use of the mathematical calculus of probability, and so could only be formulated as a theory *after* the mathematical calculus of probability had been invented. Now historians of mathematical probability normally date its first appearance to a correspondence between Pierre de Fermat and Blaise Pascal which took place in the year 1654.[3] Bacon's *Novum Organum* was published in 1620, a full thirty-four years before the very first steps in the mathematical theory of probability. Thus Bacon could not have been a Bayesian.

Conversely, it is perfectly possible to be a Bayesian but not an inductivist. Carnap seems to have held this position in his later period (cf. Carnap, 1950, pp. 192–202). Here the idea would be to try to calculate the probabilities, given evidence, of generalizations or predictions while denying that such generalizations or predictions could be obtained by the inductive method.

But despite these logically possible positions, there is still a natural connection between inductivism and Bayesianism, which means that the two theories are often held together. Thus an in-ductivist holds that generalizations and predictions can be obtained from carefully collected observations by a process of inductive inference. But once a generalization or prediction has been obtained in this way, it would seem a natural next step to try to calculate its probability given the evidence. This next step is the procedure which the Bayesians attempt to carry out. It is not surprising, therefore, to find that Russell and the Cambridge school are both inductivists and Bayesians.

Given this general outlook, it was natural for those working in mathematically minded Cambridge to investigate the interpretation of probability needed to carry out the Bayesian programme. Keynes (1921) adopted a logical interpretation of probability, and saw this probability logic or inductive logic as an extension of the deductive logic used by Russell and Whitehead in *PM*. Whereas deductive logic provided a foundation for mathematics, inductive logic was intended to provide a justification for science. Ramsey (1926) criticized Keynes's logical interpretation, and developed instead what is known as the 'subjective theory of probability'. This point of view was introduced and developed independently in Italy by the mathematician and philosopher Bruno de Finetti.

The nature of these interpretations of probability and their relevance to the Bayesian programme are fascinating questions, but to pursue them further would involve us in mathematical considerations which lie outside the scope of this non-technical book.[4] In what follows, therefore, we shall be concerned almost exclusively with inductivism, and Bayesianism will be mentioned only in passing. I will next give a brief historical account of the Vienna Circle which, in the 1920s and early 1930s, continued the work of Russell and the Cambridge school.

1.4   The Vienna Circle

The origins of the Vienna Circle go back to a group of enthusiastic research students who, in 1907, used to meet in an old Viennese coffee-house every Thursday night to discuss problems of science and philosophy.[5] Three of these young men were working in different fields (Philipp Frank in physics, Hans Hahn in mathematics, and Otto Neurath in economics), but they shared an interest in philosophy of science. Some years later, in 1921, Hahn obtained a chair of mathematics in Vienna, and, when the Mach–Boltzmann professorship of the inductive sciences became vacant the next year, Hahn exerted his influence to get Moritz Schlick appointed.

Schlick's arrival in Vienna in 1922 marked the real beginning of the Vienna Circle. Schlick in fact arranged a seminar for a small invited group, and this group came to be known as the Vienna Circle. At first the principal participants were Hahn and Neurath; and Philipp Frank, who was professor of theoretical physics in Prague, paid frequent visits. Later the group was extended to include, among others, Viktor Kraft, Herbert Feigl, Friedrich Waismann, and Kurt Gödel. Schlick was able to obtain the post of instructor at the University of Vienna for Carnap, who joined the circle in 1926 and rapidly became one of its leading figures. In 1929 the group published its manifesto: *Wissenschaftliche Weltauffassung: Der Wiener Kreis* ('The Scientific Conception of the World: The Vienna Circle'), which was written mainly by Neurath; and in 1930 it started its own journal, *Erkenntnis* ('Cognition') edited by Carnap and Reichenbach.

The manifesto contains an appendix listing the members of the Vienna Circle – fourteen in number. There is then a list of those sympathetic to the Vienna Circle, which includes Ramsey and Reichenbach. Reichenbach was in Berlin, and so was not formally a

member of the Vienna Circle, but he shared their interests and
outlook, and can be considered an associate member of the Circle.
The manifesto's appendix gives, finally, an honorific list of 'leading
representatives of the scientific world-conception'. This list contains
only three names: Albert Einstein, Bertrand Russell, and Ludwig
Wittgenstein.

Karl Menger, in his contribution to Gadol (ed.), 1982, gives a
vivid description both of his invitation to join the Vienna Circle, or
the Schlick Circle (*Schlick-Kreis*) as it was then called, and of the
meetings of the Circle. Here is his account of the invitation:

> When I returned to the University of Vienna in the fall of 1927, in
> order to teach in the chair for geometry, the mathematician Hans
> Hahn asked me whether I would like to join the *Schlick-Kreis*, the
> group that abroad became famous under the name of Vienna Circle.
> 'I attend regularly', he said, 'and so do Carnap, Neurath and a few
> younger people; and Philipp Frank visits us whenever he comes to
> Vienna [from Prague]. We meet every other Thursday evening on the
> ground floor of the wing of this building in the Boltzmanngasse.' We
> were talking in the L-shaped university building that housed the
> mathematics and physics institutes. (Menger, 1982, p. 85)

Of the meetings he writes:

> The room on the ground floor in which the Circle met – rarely more
> than 20 persons – was rather drab. We would stand in little groups
> talking until Schlick would clap his hands and we would be seated.
> Sometimes Schlick would begin by reading to us a letter that he
> had received dealing with problems that we had discussed or planned
> to consider. His correspondents included Einstein and Russell. He
> would open other sessions by reading announcements of new publi-
> cations (especially British ones) and would promise to report about
> some or ask volunteers to review them. Occasionally, Schlick would
> introduce a guest passing through Vienna. Then there would begin
> a discussion on the topic proposed in the preceding meeting or some-
> one's report about work in progress. But in none of the many
> meetings that I attended in the course of the years did the debates ever
> touch political or economic problems. Even men with strong political
> convictions never expressed them in the discussions of the Circle. It
> was in these discussions that Schlick particularly excelled both as a
> stimulating participant and as a moderator of ideal poise. (p. 86)

The philosophical views of the Vienna Circle came to be known
as 'logical positivism', though really the term 'logical empiricism',

introduced earlier, is more appropriate. As we might expect, Russell was a major influence. In his memoir of Hahn, one of the founding members of the Vienna Circle, Menger writes: 'During the early 1920s he developed a great admiration for the works of Bertrand Russell. He reviewed some of them in the *Monatshefte für Mathematik und Physik*. In one of these reviews Hahn suggested that one day Russell might well be regarded as the most important philosopher of his time' (Menger, 1980, p. xi). Hahn also conducted a seminar on Russell and Whitehead's *Principia Mathematica* in the academic year 1924–5 during which the participants went through that work chapter by chapter.

Another important influence on the Vienna Circle was Ludwig Wittgenstein, whose first important philosophical work, the *Tractatus Logico-Philosophicus*, was published in 1921. The Vienna Circle devoted itself to reading this book 'paragraph by paragraph' during the academic year 1926–7 (Menger, 1980, p. xii). Curiously enough, Wittgenstein, though in Austria, did not take part in these discussions. After finishing the *Tractatus*, he decided to give up philosophy and become a village schoolmaster in Austria. It was not until 1929 that Wittgenstein decided to resume philosophical research, and he then returned to Cambridge in England to do so. Wittgenstein never attended a meeting of the Vienna Circle, but he had occasional discussions with selected members of the Circle (notably Schlick and Waismann) in the period 1929–32. Notes of some of these conversations were made by Waismann, and are printed in *Ludwig Wittgenstein und der Wiener Kreis* (McGuinness, 1967).

Roughly speaking, we can say that Russell influenced the Vienna Circle as regards their logicism in mathematics and their interest in induction, whereas Wittgenstein influenced them as regards the question of the demarcation between science and metaphysics. Wittgenstein's *Tractatus* will therefore be discussed in some detail in Part IV, which deals with the demarcation problem. Some further information about Wittgenstein's life will also be given at that stage.

As this book develops, it will become clear that there are many objections to the logical empiricism advocated first by Bertrand Russell and then by the Vienna Circle. Indeed, logical empiricism is held by few, if any, philosophers today. Yet although many of the detailed views of the Vienna Circle have been shown to be false, a case can still be made for some of the Circle's general principles. Thus, in their manifesto we find the following: 'In the research work of all branches of empirical science this *spirit of a scientific conception of the world* is alive. However only a very few leading thinkers give it systematic thought or advocate its principles' (Neurath *et al.*, 1929, p. 3). Even if the Vienna Circle were wrong in their precise analysis of the scientific conception of the world, they may have been right that there is such a conception, and right also to advocates its principles.

1.5   The Twentieth-Century Revolution in Physics

So far we have discussed logical and philosophical influences on the Vienna Circle, but scientific influences were also of the greatest importance. The years 1900–30 were those of a great revolution in physics, which called into question the Newtonian mechanics that had been accepted for nearly two centuries, and gave birth to the new theories of relativity and quantum mechanics. The revolution got under way in the early years of the twentieth century, with the development of the special theory of relativity by Hendrik Lorentz, Henri Poincaré, and Albert Einstein,[6] and the first steps in quantum theory by Max Planck and Albert Einstein. A notable event, as already remarked, was the confirmation of Einstein's new theory of gravitation by the eclipse experiment of 1919. During the 1920s, when the Vienna Circle was being formed and was developing its 'scientific conception of the world', Werner Heisenberg, Erwin Schrödinger, and Paul Dirac were introducing the new quantum mechanics.

There were strong interactions between the philosophizing of the Vienna Circle and the great revolution in physics. On the one hand, the circle devoted much time to discussing the conceptual problems of the new physics – the nature of space and time in the light of relativity and the paradoxes of quantum mechanics. On the other hand, many of the creators of the new theories in physics adopted some form of empiricism or positivism as their philosophy of science. This was true, for example, of Einstein and Heisenberg. Admittedly, Einstein turned against empiricism in his later years, but this was after his most creative period as a physicist had finished.

There were also personal contacts between members of the Vienna Circle and the leading physicists of the time. Schlick was a good friend of Einstein's, and the two of them engaged in a considerable correspondence concerning the philosophical interpretation of relativity. Einstein even assisted Schlick in his academic career, helping Schlick to obtain a professorship in Kiel in 1921, the year before Schlick moved to Vienna. (For an excellent account of the Einstein–Schlick correspondence, see Howard, 1984.)

As this book is non-technical in character, we will not pursue the more technical issues in the philosophy of physics, interesting though these are. It is worth noting, however, how the revolution in physics influenced both the Vienna Circle's choice of fundamental philosophical problems and the way in which they pursued those problems. The most important factor was the change in attitude towards Newton's theory of mechanics and gravitation. Prior to the revolution, this theory had been regarded by most scientists as virtually, if not entirely, certain. Newton's theory had indeed had a most remarkable run of empirical success, explaining a mass of observations and successfully passing the experimental tests to which it had been subjected. Yet the new theories of relativity and quantum mechanics showed that Newton's theory was at best an approximation, and actually gave incorrect results in many circumstances – for example, for bodies moving with very high velocities, for bodies near very large gravitating masses, and in the micro-world. The failure of such an important and previously well-confirmed scientific theory gave a new significance to Hume's philosophical doubts about induction. The problem was posed as to whether, and to what extent, scientific theories could be justified inductively by observational and experimental evidence.

1.6   Popper

Let us now consider an important thinker who, though not a member of the Vienna Circle, had close links with it. Karl Popper was born in 1902, so was some years younger than Wittgenstein and the founder members of the Vienna Circle. He was never invited to attend the Vienna Circle's seminar, and was wittily, but accurately, characterized by Neurath as the Circle's 'official opposition'. Popper's book *The Logic of Scientific Discovery* (1934) was first published in the Vienna Circle's series edited by Schlick. It deals with the topics most characteristic of the Vienna Circle – induction, demarcation, probability and confirmation, the conceptual problems of quantum mechanics, and so on. Thus Popper shared with the Vienna Circle an interest in the same set of problems. The difference

lay in the answers he gave to these questions. Popper criticized
the views of the Vienna Circle on most fundamental philosophical
issues, and he developed philosophical theories which differed profoundly from theirs.

It cannot be said, however, that Popper differed from the Vienna
Circle on every topic. For example, in *The Logic of Scientific Discovery*, Popper argues for a version of von Mises' frequency theory
of probability. Richard von Mises, like Reichenbach, was an associate
of the Vienna Circle, and the frequency approach to probability,
which he and Reichenbach adopted, was accepted by many members of the Circle. Popper also shared the Vienna Circle's admiration
for Einstein and Russell. Thus, in his *Quantum Theory and the Schism
in Physics*, Popper speaks of 'my boundless admiration for the work
of Einstein' (1982b, p. 157), and, in his *Realism and the Aim of
Science*, writes: 'In the autumn of 1935 . . . I was taken to a meeting
of the Aristotelian Society. Bertrand Russell, whom I had long
admired as the greatest philosopher since Kant, read a paper on
"The Limits of Empiricism"' (1983, p. 12). But Popper did not, as
we shall see, share the Vienna Circle's admiration for Wittgenstein.

All his life Popper has remained an essentially oppositional figure.
Not long after finishing *The Logic of Scientific Discovery*, he left
Vienna to take up a lectureship in philosophy in New Zealand.
Since he did not find there any major school of philosophy to
attack, he devoted his time to writing a long criticism of Plato and
Marx, whom he characterized as 'enemies of the open society'.
After the war, Popper went to England, and once again found
himself in largely isolated opposition to the dominant philosophical
trend. Wittgenstein had returned to Cambridge in the 1930s, and
he, together with important figures in Oxford (Austin and Ryle),
created a system of thought known as 'linguistic philosophy'.
After the Second World War, this became the standard approach to
philosophy in Britain, and Popper was one of the few major figures
in the country to reject linguistic philosophy virtually *in toto*. Popper
had a small but lively school in London, and exercised considerable
influence on the development of philosophy of science world-wide.
Although he had few followers in the British philosophical community, his views had, and still have, many supporters among
scientists.

So for a time after the Second World War, Popper was in London,
separated by only a few miles from Wittgenstein in Cambridge. Yet
the two philosophers seem only to have had one encounter. Already
legends have sprung up about this event. The details differ, but

most versions involve Wittgenstein brandishing a poker ferociously and then storming from the room in a rage. Popper has provided his own account of the meeting in his autobiography, *Unended Quest* (1976, pp. 122–4). In 1946 Popper received an invitation to speak to the Moral Sciences Club at Cambridge, giving his talk on 26 October of that year. In reference to this, Popper explains his 'custom whenever I am invited to speak in some place, of trying to develop some consequences of my views which I expect to be unacceptable to that particular audience' (p. 124). This 'custom' certainly reveals a great deal about Popper's attitudes. On this occasion, it led him to speak on the subject 'Are there Philosophical Problems?' and to argue that there are genuine, important such problems. This was a direct challenge to Wittgenstein's view, which will be discussed in Part IV, that philosophical problems are all pseudo-problems arising from the misunderstanding of language. Popper recounts Wittgenstein's very negative reaction to the paper. His description ends as follows:

> At that point Wittgenstein, who was sitting near the fire and had been nervously playing with the poker, which he sometimes used like a conductor's baton to emphasize his assertions, challenged me: 'Give an example of a moral rule!' I replied: 'Not to threaten visiting lecturers with pokers.' Whereupon Wittgenstein, in a rage, threw the poker down and stormed out of the room, banging the door behind him. (p. 123)

Richard Braithwaite, in whose rooms in King's College, Cambridge, the meeting took place, used to explain in after years that Wittgenstein was merely poking the fire – violently!

In Part IV I will argue, somewhat paradoxically in view of the strained relations between the two philosophers, that Wittgenstein's later ideas on language and meaning actually lend support to Popper's account of metaphysics.

1.7   The Dispersal of the Vienna Circle

We must now briefly describe the melancholy events which led to the expulsion of the Vienna Circle and its 'official opposition' from their native city. In 1934, the year of the publication of *The Logic of Scientific Discovery*, the Austrian Fascist Party, under Dollfuss, seized power. Parliament was suspended, and socialists disbanded and jailed. But Dollfuss himself was murdered by the Nazis in July 1934. None the less independent Austrian Fascism continued for a while with Schuschnigg as the new Führer of the Fatherland Front. But long it could not be, and on 12 March 1938 Hitler invaded and occupied Austria.

Many members of the Vienna Circle were Jewish, and even those who were 'Arian' held liberal or socialist opinions unlikely to appeal to Fascists and Nazis. Moreover, the Vienna Circle's 'scientific conception of the world' was a threat to the highly unscientific racial theories which were a key element in Nazi propaganda.

The most left-wing member of the Vienna Circle was Otto Neurath. Although not Marxist in any orthodox sense, he had considerable sympathy for Marxism, and was in charge of central planning in the temporary Spartacist government set up in Bavaria after the first world war. When the Spartacists were overthrown, Neurath was sentenced to imprisonment, but was released on the intervention of the Austrian government. He would almost certainly have been imprisoned by the Fascists when they seized power in 1934, but, luckily, was in Moscow at the time explaining his system of international communication using pictograms. Instead of returning to Vienna, he went to Holland, but had to flee again when the Nazis invaded in 1940. He then went to Oxford, where he remained till his death in 1944.

The next dramatic event was the assassination of Schlick, which took place in 1936. Schlick had examined (and failed) a thesis on ethics written by a Nazi student (Nelbock). Later, when ascending the philosophers' stairway in the university building, he was shot dead by this student. The student was sent to prison for a mere ten years, though he could have been sentenced to death, and was then released by the Nazis when they occupied Vienna in 1938. Nelbock applied for a full pardon in 1941 on the grounds that he had done a useful service in disposing of a Jewish professor. (In fact, Schlick was not Jewish, but was descended from the Prussian nobility.) Nelbock was given a post in the geological division of the mineral oil administration of the war economy, where he worked till the end of the war. He died in 1954.

After Schlick's assassination, most remaining members or associates of the Vienna Circle left in a hurry. This brilliant group of philosophers had, like Dante, to eat the bitter bread of exile.[7] But the loss to the culture and civilization of Austria and the German-speaking world was, at the same time, a gain to the English-speaking world. The displaced philosophers, scientists, and mathematicians settled in the USA, Britain, and the Commonwealth, where they exerted a great influence on the development of philosophy of science. This process was reinforced by some British and American philosophers who had studied with the Vienna Circle in Austria and then returned home. This course was followed by Willard van Orman Quine from Harvard and Alfred Jules Ayer from Oxford. Ayer's version of logical empiricism, published in 1936 in his book *Language, Truth, and Logic*, spread Vienna Circle ideas in Britain. Very interesting too are Ayer's criticisms and developments of his original views which appeared as an introduction to the second (1946) edition of his book. I will discuss some of these later on.

I have now introduced most of the philosophers whose views we shall consider in this book. There remain two French thinkers (Pierre Duhem and Henri Poincaré) who flourished before the first world war and whose work in the philosophy of science is of the highest importance. We have already encountered Poincaré as one of the men involved in the development of special relativity; but, in addition to his work in mathematics and physics, Poincaré wrote books of the greatest significance on the philosophy of science.

It will be convenient to give biographical notes on Duhem and Poincaré in chapter 3. For the moment we have enough members of the cast to continue our philosophical drama, and what better way to do so than to consider Popper's criticisms of the inductivism of Russell and the Vienna Circle.

2

Popper's Critique of Inductivism. His Theory of Conjectures and Refutations (or Falsificationism)

2.1 Popper's Critique of Inductivism

The inductivist thinks that science proceeds by first collecting observations or data (Bacon's 'countless grapes . . . ripe and fully seasoned') and then inferring laws and predictions from this data by induction. Popper argues against this that one cannot simply observe without a theoretical background. Here is how he puts the argument:

> The belief that science proceeds from observation to theory is still so widely and so firmly held that my denial of it is often met with incredulity. I have even been suspected of being insincere – of denying what nobody in his senses can doubt.
>
> But in fact the belief that we can start with pure observations alone, without anything in the nature of a theory, is absurd; as may be illustrated by the story of the man who dedicated his life to natural science, wrote down everything he could observe, and bequeathed his priceless collection of observations to the Royal Society to be used as inductive evidence. This story should show us that though beetles may profitably be collected, observations may not.
>
> Twenty-five years ago I tried to bring home the same point to a group of physics students in Vienna by beginning a lecture with the following instructions: 'Take pencil and paper; carefully observe, and write down what you have observed!' They asked, of course, *what* I wanted them to observe. Clearly the instruction, 'Observe!' is absurd. . . . Observation is always selective. It needs a chosen object, a definite task, an interest, a point of view, a problem. And its description presupposes a descriptive language, with property words; it presupposes similarity and classification, which in its turn presupposes interests, points of view, and problems. (Popper, 1963, p. 46)

The same thing applies even if we go right back to the beginnings of science or of an individual human life. Popper argues that something like modern science developed in ancient Greece through criticism and modification of an older mythological picture of the world. New-born babies do not have blank minds, but inborn expectations as the result of genetic inheritance. However, as Popper points out, these expectations may be disappointed. The new-born child expects to be fed, but may be abandoned and starve.

Let us see how this argument of Popper's applies to the Kepler example. It will be remembered that Tycho Brahe made his observations of the heavens, particularly of the planets, between 1576 and 1597. Now Copernicus's new theory of the universe had been published in 1543. By the 1570s there was a major theoretical dispute in astronomy between the upholders of the older Aristotelian-Ptolemaic view that the Earth was stationary at the centre of the universe and the Sun went round the Earth and the Copernicans, who thought that the Sun was stationary at the centre and the Earth went round the Sun. Tycho Brahe's observations were relevant to this theoretical controversy. At a more basic level, even his division of heavenly bodies into stars and planets involved a theoretical classification. Indeed, this classification was different in the two contending theories. In the Ptolemaic theory, a planet was a heavenly body which was not a fixed star and which moved round the Earth; so the Sun was a planet, but the Earth was not a planet. In the Copernican theory, a planet was a body which was not a fixed star and which moved round the Sun; so, on this account, the Sun was not a planet, but the Earth was.

Let us turn to a second argument of Popper's. This concerns the principle of induction. Some inductivists think that their inductive inferences are justified by the principle of induction. But how is the principle of induction itself justified? If we try to justify it inductively by experience, we get into a circle, since any induction from experience depends on the principle of induction. We might try to avoid making this circle vicious by justifying the basic principle of induction by a higher-order principle of induction. But then this would, in turn, have to be justified by a still higher-order principle of induction, and so on. Thus, if we try to justify the principle of induction inductively by experience, we get either a vicious circle or an infinite regress. The other possibility is to try to justify it independently of experience, or *a priori*. However, this looks like a blind act of faith. This is how Popper puts the argument:

Inconsistencies may easily arise in connection with the principle of induction. . . . For the principle of induction must be a universal statement in its turn. Thus if we try to regard its truth as known from experience, then the very same problems which occasioned its introduction will arise all over again. To justify it, we should have to employ inductive inferences; and to justify these we should have to assume an inductive principle of a higher order; and so on. Thus the attempt to base the principle of induction on experience breaks down, since it must lead to an infinite regress.

Kant tried to force his way out of this difficulty by taking the principle of induction (which he formulated as the 'principle of universal causation') to be 'a priori valid'. But I do not think that his ingenious attempt to provide an a priori justification for synthetic statements was successful.

My own view is that the various difficulties of inductive logic here sketched are insurmountable. (1934, p. 29)

Popper (1963, p. 289) argues that the same objection applies to the principle of uniformity of nature, which he regards as a kind of principle of induction.

It is interesting to see how Russell responded to objections of this sort. Russell agreed that we cannot justify the principle of induction by experience, and concluded that we must accept the principle a priori, or, as he put it, 'on the ground of its intrinsic evidence'. This is what Russell said: 'We can never use experience to prove the inductive principle without begging the question. Thus we must either accept the inductive principle on the ground of its intrinsic evidence, or forgo all justification of our expectations about the future' (Russell, 1912, p. 106). He regards forgoing all justification of our expectations about the future – that is, complete scepticism – as nothing less than intellectually frivolous. Thus he advocates an a priori acceptance of the principle of induction 'on the ground of its intrinsic evidence'. He thinks that the acceptance of the principle of induction is necessary in order to be a scientist: 'The general principles of science, such as the belief in the reign of law, and the belief that every event must have a cause, are as completely dependent upon the inductive principle as are the beliefs of daily life' (p. 107).

Russell thus holds that we must, however reluctantly, believe in the principle of induction as a kind of blind act of faith in order to do science. And here we come to Popper's most fundamental criticism; for Popper thinks that we can be scientists and do science without making any inductive inferences. Hence we do not need a principle of induction to justify inductive inferences; consequently, there is no need to have blind faith in such a principle. Popper thus gets round the problem by suggesting a non-inductive theory of scientific method. This is his method of conjectures and refutations, which I will now expound.

2.2   Popper's Theory of Conjectures and Refutations (or Falsificationism)

Hume pointed out that, from observations and deductive logic, we can never infer the truth of a generalization. Thus however many white swans we see, we can never infer the truth that all swans are white.

Popper, however, observed that, although observations and deductive logic cannot establish the *truth* of a scientific generalization (or verify it), they can establish its *falsity* (or refute or falsify it). Thus, from the observation 'This is a black swan', we can infer by deductive logic that the generalization 'All swans are white' is false. In other words, we can refute or falsify a scientific generalization. Popper refers to this as the asymmetry between falsification and verification.

This leads Popper to his *conjectures and refutations*, or *falsificationist*, account of scientific method. Science does not start with observations, as the inductivist claims, but with conjectures. The scientist then tries to refute these conjectures by criticism and testing (experiments and observations). A conjecture which has withstood a number of severe tests may be tentatively accepted, but only tentatively. We can never know a scientific theory, law, or generalization with certainty. It may break down on the very next test or observation (as in the case of the discovery of black swans in Australia).

To illustrate the tentative, conjectural nature of scientific knowledge, Popper is fond of citing the example of Newtonian mechanics. Newton's theory produced an extraordinarily good fit with observation and experiment from the time it was published (1687) until 1900. Nevertheless, between 1900 and 1920 it was found to be inaccurate in some respects, and corrections were introduced using relativistic mechanics.

2.3   The Distinction between Discovery and Justification

Popper's theory of conjectures and refutations leads naturally to a distinction between the *discovery* of scientific hypotheses and their *justification* or *validation*. This is how Popper himself puts the matter:

> The work of the scientist consists in putting forward and testing theories.
>
> The initial stage, the act of conceiving or inventing a theory, seems to me neither to call for logical analysis nor to be susceptible of it. The question how it happens that a new idea occurs to a man – whether it is a musical theme, a dramatic conflict, or a scientific theory – may be of great interest to empirical psychology; but it is irrelevant to the logical analysis of scientific knowledge. This latter is concerned not with *questions of fact* (Kant's *quid facti?*), but only with questions of *justification or validity* (Kant's *quid juris?*). Its questions are of the following kind. Can a statement be justified? And if so, how? Is it testable? Is it logically dependent on certain other statements? Or does it perhaps contradict them? (1934, p. 31)

Note that this theory of Popper's brings about a certain *rapprochement* between science and the arts. Great scientists can, according to Popper, possess the kind of creativity which is recognized as a possession of great artists. A classic example of creative intuition in science is provided by Kekulé's discovery in 1865 that the six carbon atoms of the benzene molecule are arranged in a ring. Kekulé, while meditating on the structure of benzene, fell asleep and dreamt of a snake biting its tail. On awakening, he hit on the solution of his problem.

Popper goes on to consider the view that philosophers of science should attempt to give a rational reconstruction of the process of discovery. As he puts it: 'Some might object that it would be more to the purpose to regard it as the business of epistemology to produce what has been called a "*rational reconstruction*" of the steps that have led the scientist to a discovery – to the finding of some new truth' (p. 31). Popper, however, rejects such a view, because he thinks that discovery always contains an irrational, creative element. As he puts it:

> My view of the matter, for what it is worth, is that there is no such thing as a logical method of having new ideas, or a logical reconstruction of this process. My view may be expressed by saying that every discovery contains 'an irrational element', or 'a creative

intuition', in Bergson's sense. In a similar way, Einstein speaks of the 'search for those highly universal laws . . . from which a picture of the world can be obtained by pure deduction. There is no logical path, he says, 'leading to these . . . laws. They can only be reached by intuition, based upon something like an intellectual love (*Einfühlung*) of the objects of experience. (p. 32)

So Popper's view is that there is no such thing as a logic of scientific *discovery* – only a logic of scientific *testing*.

The distinction between *discovery* and *justification* is related to the distinction between *inductivism* and *Bayesianism* which we made in the last chapter. Bayesianism is indeed a theory of justification, not of discovery. Bayesians seek to justify scientific generalizations or predictions by showing that, although they are not certain, they can none the less be shown to be probable, given the evidence used to support them. It is thus possible to be a Bayesian regarding the justification of scientific generalizations and predictions, while denying the inductivist theory of how they are discovered. Indeed, as we remarked in the last chapter, Carnap held a position of this sort in his later period. Thus, in his 1950 book, he accepts, with acknowledgement, the Einstein–Popper position on discovery. As he puts it:

> But in one point the present opinions of most philosophers and scientists seem to agree, namely, that the inductive procedure is not, so to speak, a mechanical procedure prescribed by fixed rules. If, for instance, a report of observational results is given, and we want to find a hypothesis which is well confirmed and furnishes a good explanation for the events observed, then there is no set of fixed rules which would lead us automatically to the best hypothesis or even a good one. It is a matter of ingenuity and luck for the scientist to hit upon a suitable hypothesis; . . . . This point, the impossibility of an automatic inductive procedure, has been especially emphasized, among others by Karl Popper . . . , who also quotes a statement by Einstein . . . . The same point has sometimes been formulated by saying that it is not possible to construct an inductive machine. The latter is presumably meant as a mechanical contrivance which, when fed an observational report, would furnish a suitable hypothesis, just as a computing machine when supplied with two factors furnishes their product. I am completely in agreement that an inductive machine of *this* kind is not possible. (Carnap, 1950, p. 192)

The development since 1950 of ever more powerful computers has reawakened hopes that it might after all be possible to construct an automatic inductive machine. Indeed, a branch of the subject of artificial intelligence, known as *machine learning*, has as its aim the programming of computers to produce generalizations when fed with data. In the next chapter we shall examine one approach to machine learning which has been developed by Herbert Simon and his team at Carnegie–Mellon University.

Returning to Carnap in the 1950s, we must next emphasize that, although he rejected an inductivist, or Baconian, account of scientific discovery, he accepted the Bayesian theory of scientific justification. He believed that scientific predictions could be justified inductively by showing that they have a high probability on the known evidence. Popper, on the other hand, has always rejected both inductivism as a theory of discovery and Bayesianism as a theory of justification.

2.4   Some General Observations on Popper's Theory of Scientific Method

Popper gives the following summary of his theory of scientific method:

> Knowledge can grow, and . . . science can progress – just because we can learn from our mistakes.
>
> The way in which knowledge progresses, and especially our scientific knowledge, is by unjustified (and unjustifiable) anticipations, by guesses, by tentative solutions to our problems, by *conjectures*. These conjectures are controlled by criticism; that is, by attempted *refutations*, which include severely critical tests. They may survive these tests; but they can never be positively justified: they can neither be established as certainly true nor even as 'probable' (in the sense of the probability calculus). (1963, Preface, p. vii)

This is a very interesting passage, and I will make a number of comments on it. To begin with, Popper speaks of our knowledge progressing 'by unjustified . . . anticipations'. A learned reader might suspect that there is here a hidden reference to Bacon, and a desire to make what Bacon regarded as undesirable into an integral part of scientific procedure. These suspicions would be quite correct, for, to a passage in his earlier *Logic of Scientific Discovery* (1934), Popper adds a footnote referring to a section of the *Novum Organum*

(First Book, XXVI), which was quoted in 1.1. The passage in question runs as follows:

> Like Bacon, we might describe our own contemporary science – 'the method of reasoning which men now ordinarily apply to nature' – as consisting of 'anticipations, rash and premature' and of 'prejudices'.
>
> But these marvellously imaginative and bold conjectures or 'anticipations' of ours are carefully and soberly controlled by systematic tests. Once put forward, none of our 'anticipations' are dogmatically upheld. Our method of research is not to defend them, in order to prove how right we were. On the contrary, we try to overthrow them. Using all the weapons of our logical, mathematical, and technical armoury, we try to prove that our anticipations were false – in order to put forward, in their stead, new unjustified, and unjustifiable anticipations. . . . (Popper, 1934, 278–9)

It should next be observed that at the end of the passage quoted from the preface to his 1963 book, Popper explicitly rejects Bayesianism. He argues that 'these conjectures . . . can neither be established as certainly true nor even as "probable" (in the sense of the probability calculus).' Of course, the Bayesian thesis is precisely that scientific conjectures can be made probable in the sense of the mathematical calculus of probability. However, Popper is not content just to criticize the Bayesian attempt to justify scientific conjectures; he states the much stronger thesis that such conjectures cannot be justified at all, or, as he puts: 'these conjectures . . . can never be positively justified'. He also speaks in both the 1934 and the 1963 passages of 'unjustified (*and unjustifiable*) anticipations' (my emphasis). It should be clear from all this that Popper's critique of inductivism contains quite a number of different theses, of different kinds and of different strengths. I will next try to disentangle a few of them, and to comment on their plausibility.

Let us take first Popper's thesis that there are no such things as inductive inferences, analogous to deductive inferences, by which scientific generalizations and predictions can be obtained from observational data. Popper suggests instead that all such generalizations and predictions are conjectures, and that the important point is *not* how such conjectures are obtained – any method will do – but that they should be tested severely once they have been proposed.

This thesis of Popper's seems to me plausible, and it also brings about a great simplification in the theory of scientific method. We do not have to postulate any curious process of inductive inference

and investigate its character. Instead, the simple procedure of con-
jecturing, followed by deductive inferences, suffices. Moreover, the
simplification of scientific method does not end here.

As we have seen, those who, like Russell, postulate inductive
inferences naturally raise the question of how such inferences can be
justified. This leads them to claim that inductive inferences need to
be justified by some *principle of induction*, or *principle of the uniformity
of nature*. It has to be said, however, that this whole approach is
most unsatisfactory. As Popper points out, the justification of these
principles in turn is liable to lead to a vicious circle or an infinite
regress. Russell's attempt to avoid this by saying that we must
accept these principles *a priori* on the ground of their intrinsic evidence is hardly very plausible. As a matter of fact, Russell's own
formulation of the principle of induction contains an error, as we
saw, and even when this error is corrected, it is by no means clear
that the resulting principle is correct. Matters are no better with the
principle of the uniformity of nature, which Russell formulates as
follows: 'The belief in the uniformity of nature is the belief that
everything that has happened or will happen is an instance of some
general law to which there are *no* exceptions' (1912, p. 99).

There seems more reason for believing this to be false than true,
however. Is it not more likely that some things happen by chance
and are not instances of general laws? The principle does not appear
to be necessary for science either. Surely, science would still be
possible, even if the universe contained a certain amount of intrinsic
randomness. To sum up then: it seems to be difficult, if not impossible, to formulate the alleged principles of induction and
uniformity of nature in such a way that they are even plausible, let
alone obviously true *a priori*. Surely, then, it is better to get rid of
these obscure and unsatisfactory principles if we can. Popper's first
thesis shows a way in which this can indeed be done. It is thus
marks, in my view, a definite advance over the Cambridge school.

Popper's second thesis is that Bayesianism should be rejected.
Unfortunately we cannot discuss the arguments for and against this
view in the present, non-technical book. It can be observed in
general terms, however, that this thesis, like the first, is by no
means implausible. The Bayesian claims to be able to calculate the
probability of some scientific prediction, given the evidence in its
favour. Can such computations really be performed? Or do we have
here a misuse of the mathematical theory of probability? Popper's
scepticism about such calculations seems *prima facie* quite reasonable.

But, while Popper's first two theses are quite plausible, the same cannot be said of his third thesis, the thesis that scientific conjectures can never be positively justified. Consider a theory (T, say) put forward at time t_1 by a scientist (Dr E, say). Let us suppose that at t_1 there is really no evidence in favour of T, so that it can be regarded as purely conjectural. Between t_1 and t_2, however, Dr E and others show that T can explain a whole mass of observational data. T is, moreover, subjected to a whole series of experimental tests, and by t_2 has passed every single one of them. Now most people would surely say that, while there was no evidential justification for T at time t_1, the evidence which has accumulated by time t_2 has strongly vindicated T, and that any technologist would by then be justified in using T as the basis for some practical application. But Popper in his third thesis seems committed to the view that T is just as unjustified at t_2 as it was at t_1. Indeed, a theory like T cannot be justified, because such theories ('anticipations') are intrinsically unjustifiable. In short, Popper's third thesis flies in the face of common sense, and does not seem to me to be acceptable. Admittedly, those who deny Popper's third thesis have to explain how exactly scientific conjectures can come to be justified by the evidence used to support them, and this is certainly no easy matter. An investigation of the matter would once again involve us in mathematical questions about probability, so cannot be attempted here. But the point I would like to stress by way of conclusion is that it is perfectly possible to accept Popper's first two quite plausible theses without accepting his rather implausible third thesis.

Let us now turn from Popper's critique of inductivism to his own positive view that scientific knowledge grows through conjectures and refutations. Now it would seem to be a good plan to apply Popper's own methodological principles to his theory of scientific method and to subject that theory to the most severe tests we can devise. Any theory of scientific method can be tested against episodes from the history of science. Let us therefore consider some famous advances in science which appear at first sight to fit the inductive model much better than Popper's falsificationist model. We can then see whether it might be possible to account for these scientific developments in terms of conjectures and refutations. I have already introduced one such example: namely, Kepler's discovery that planets move in ellipses, and we shall begin a more detailed consideration of this example in the next section.

Kepler's achievement belongs to astronomy and physics, and was carried out in the seventeenth century; but we must beware of basing any account of scientific method too much on examples

drawn from a single branch of natural science or from a single
historical period of scientific development. I will therefore supplement the Kepler example with two further examples drawn from
twentieth-century science, and from the biological and medical field.
The first of these is Alexander Fleming's discovery of penicillin.
I will recount the story in more detail later, but its outlines are
well known. Fleming observed a culture-plate which had become
accidentally contaminated by a mould. His own photograph of
the culture-plate is reproduced as Plate 1. Fleming concluded that
the mould was producing a substance which had destroyed the
bacteria growing on the culture-plate and which might, therefore,
be a suitable antibiotic for the treatment of bacterial infections.
Surely, here, observation preceded theory, and Fleming made an
inductive inference from his observation. My final example is
closely related to Fleming's discovery, but has some rather different
features. It is the discovery of the sulphonamide drugs by the
German chemical company I. G. Farben Industrie. The chemists
working for the company examined literally hundreds of chemical
compounds, until they hit, to some extent by accident, on one
which cured mice infected by haemolytic streptococci. This really
seems to fit the Baconian model of a mass of careful observations
which eventually reveal a 'secret of excellent use'. These three
examples, therefore, appear at first sight to fit the inductivist model
much better than the falsificationist model. But our further examination of them will show that this initial impression is to some
extent misleading, and that an analysis in terms of conjectures and
refutations is, in each case, more plausible than might at first
be thought. Yet the apparently inductivist features of the three
examples are not illusory, and our analysis will lead us eventually to
a kind of synthesis between inductivism and falsificationism.

2.5   Kepler's Discovery of the Elliptic Orbits of the Planets

Let us consider, then, Kepler's discovery that the planet Mars moves
in an ellipse with the Sun at one focus.[1] At first sight it seems
as if Kepler inferred his law inductively from Tycho Brahe's observations. But let us look a little more closely at what happened.

   To begin with, Kepler started his investigation with a number of
theoretical assumptions. He was already a convinced Copernican,
and so related Mars's orbit to the Sun rather than the Earth. Indeed,

he assumed that the Sun was a centre of force which governed the motions of the planets. If Kepler had tried to relate Mars to the Earth rather than to the Sun, he would never have found the elliptic orbit. He also began with the theoretical assumption that the motion of heavenly bodies was either circular or composed of a small number of circular motions. This assumption (which Kepler later rejected) had been made by astronomers since the time of Plato and Aristotle.

With these background assumptions, Kepler formulated his *first hypothesis*:

> The orbit of Mars is a circle round a centre C somewhat displaced from the sun S, and its motion is uniform with respect to a point U.

On this hypothesis (illustrated in figure 2.1) the planet moves faster when nearer the Sun, in conformity with the idea that the Sun is a centre of force influencing its motion.

Kepler made 900 folio pages in small handwriting of draft calculations relating to this hypothesis. He based it on four observed positions of Mars in opposition, and it agreed within two minutes of arc with another ten oppositions. But he then went on to test the hypothesis against some further observations of Tycho's, and this produced a deviation of 8'. This led Kepler to reject his first hypothesis.

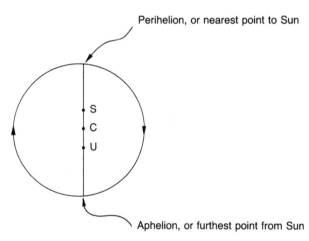

Perihelion, or nearest point to Sun

S
C
U

Aphelion, or furthest point from Sun

Figure 2.1   Kepler's first hypothesis

Kepler next tried (second hypothesis) to compose the motion out of two circles. Mars moved round the second circle (or *epicycle*), whose centre moved round the first circle (or *deferent*). The combination of deferent and epicycle produced an egg-shaped curve with the pointed end at the perihelion, the nearest point to the Sun, and the broad end at the aphelion, the furthest point from the Sun. This again could not be made to fit Tycho's data.

Kepler next tried a third hypothesis, which is mathematically equivalent to an ellipse. However, through a mathematical mistake, he got a curve which did not fit. He then tried an ellipse (fourth hypothesis) which worked, and subsequently realized that the third and fourth hypotheses were equivalent.

A closer examination of Kepler's work shows a pattern of conjectures and refutations in accordance with Popper's model. But two difficulties should be noted:

1 The assumption about the circular motion of heavenly bodies could not by itself be directly refuted. Kepler was able to refute only specific hypotheses based on it, such as his first and second hypotheses.
2 The data which Kepler used to test his hypotheses were collected by Tycho Brahe before these hypotheses had been formulated. In this sense, observation did precede theory.

Let us say a little more about these two points in turn. In connection with the first, we should make some mention of the views of Thomas Kuhn, a distinguished American philosopher of science well known for his analysis of scientific revolutions. In his famous 1962 book, Kuhn divides scientific development into revolutionary periods and non-revolutionary periods, calling the latter 'normal science'. He thinks that Popper's theory applies to revolutionary science, but not to normal science. As he says: 'I suggest then that Sir Karl has characterised the entire scientific enterprise in terms that apply only to its occasional revolutionary parts' (1970, p. 6).

As a matter of fact, however, it is questionable whether Popper's theory of conjectures and refutations gives a satisfactory account of scientific revolutions. In general, scientific revolutions involve the overthrow of some high-level theoretical assumption and its replacement by a new assumption. The difficulty is that such high-level theoretical assumptions cannot be directly refuted by observation. We can only directly refute specific hypotheses based on those assumptions. Thus a creative scientist is never forced by refutation to give up a high-level theoretical assumption. Kepler need never have abandoned the age-old assumption of the circularity of the heavenly motions in order to introduce his ellipse. In the face of the refutation of his first two specific hypotheses, he could simply have added more and more epicycles until a better fit was obtained.

A sequence of conjectures and refutations of specific hypotheses need never be revolutionary in character. Moreover, such sequences are often found to be the fine structure of what Kuhn calls 'normal science'.

Turning now to our second point, it seems undeniable that observations do sometimes precede theory (at least in the sense in which Tycho Brahe's observations preceded Kepler's theorizing). But how do such cases differ from the obviously valueless random observations which Popper parodies? We have already given part of the answer to this question. Tycho Brahe's observations were made at a time when a controversy was raging between the Ptolemaic and the Copernican theories of the world, and his observations were clearly relevant to this controversy. Another point worth noting is that Tycho Brahe's observations were also relevant to a practical problem. Trade and shipping were developing rapidly at the time, and brought with them the demand for better astronomical tables which could be used in navigation. It is significant that the King of Denmark, who financed Tycho Brahe's observatory, derived a large part of his royal income from tolls on ships passing through the Danish sound, which connects the Baltic to the North Sea.

Kepler put his astronomical discoveries to practical use, employing them to compute a new set of astronomical tables. These were called the *Rudolfine Tables* after his patron King Rudolph II, and were issued in 1627. They achieved new levels of accuracy, completely superseding all previous astronomical tables, and were at once used for navigational purposes.

Thus we may say that it is only worth collecting observations if these Baconian 'grapes . . . ripe and fully seasoned' are clearly relevant to a theoretical controversy or a practical application.

2.6   Fleming's Discovery of Penicillin: Creative Induction

The research work which led to the discovery of penicillin began because Fleming had been invited to contribute a section on the staphylococcus group of bacteria for the nine-volume *A System of*

Bacteriology which was being produced by the Medical Research Council.[2] Fleming's contribution did indeed appear in the second volume in 1929. Staphylococci are spherical bacteria which are responsible for a variety of infections. In particular, the golden-coloured *Staphylococcus aureus* is the common cause of boils, carbuncles, and other skin infections. While reading the literature on staphylococci, Fleming came across an article by Bigger, Boland, and O'Meara of Trinity College, Dublin, in which it was suggested that colour changes took place if cultures of staphylococci were kept at room temperature for several days. This interested Fleming, because the colour of a staphylococcus can be an indicator of its virulence in causing disease. He therefore decided to carry out an experimental investigation of the matter with the help of D. M. Pryce, a research scholar.

The staphylococci were cultured in glass dishes, usually 85 mm in diameter, known as *Petri dishes*. These dishes were filled with a thin layer of a gelatinous substance called *agar* to which enough nutrients could be added to allow the microbes to multiply. Using a platinum wire, some staphylococci were spread across the surface of the agar, and the plate was then incubated at a suitable temperature (usually 37°C), to allow the microbes to multiply. After this period of incubation, the dish was set aside on the bench, and was examined every few days to see if changes in the colour of some of the staphylococci could be observed.

While this fairly routine investigation was continuing, Pryce left the laboratory in February 1928 to start another job, but Fleming continued the work on his own throughout the summer. At the end of July Fleming went off for his usual summer holiday, leaving a number of culture-plates piled at the end of the bench where they would be out of the sunlight. Early in September (probably on 3 September) when Fleming had returned from his holiday, Pryce dropped in to see him. Pryce found Fleming sorting out the pile of plates on his bench. Discarded plates were put in a shallow tray containing the antiseptic lysol. This would kill the bacteria, and make the Petri dishes safe for the technicians to wash and prepare for use again. Fleming's tray was piled so high with dishes that some of them were protruding above the level of the lysol. Fleming started complaining about the amount of work he had had to do since Pryce had left him. He then selected a few of the dishes to show to Pryce. More or less by chance he picked up one in the tray of discards but above the level of the lysol. According to Pryce's later recollection, Fleming looked at the plate for a while, and then said: 'That's funny.' The plate was in fact the famous penicillin plate.

This is how Fleming himself described what happened in the paper he published in June 1929:

> While working with staphylococcus variants a number of culture-plates were set aside on the laboratory bench and examined from time to time. In the examinations these plates were necessarily exposed to the air and they became contaminated with various micro-organisms. It was noticed that around a large colony of a contaminating mould the staphylococcus colonies became transparent and were obviously undergoing lysis. (p. 226)

Fleming's photograph of the original penicillin plate is reproduced as Plate 1, and it is easy to follow his description when examining the photograph. The colonies of staphylococci are the small circular blobs, and the contaminating mould is very obvious. Near the mould the staphylococci become transparent or disappear altogether. They are obviously, as Fleming says, undergoing *lysis*, which means the dissolution of cells or bacteria. From his observation of the plate, Fleming inferred that the mould was producing a substance capable of dissolving bacteria. The mould was identified as being a *Penicillium*. At first it was incorrectly classified as *Penicillium rubrum*, but later it was found to be the much rarer species *Penicillium notatum*. Fleming accordingly gave the name *penicillin* to the bacteriolytic substance which he thought was being produced by the mould.

The events described so far may make it look as if Fleming's discovery was simply a matter of luck. Indeed, there is no doubt that a lot of luck was involved. Hare subsequently tried to reproduce a plate similar to Fleming's original one, and found to his surprise that it was quite difficult (see Hare, 1970, pp. 54–87). The general effect of Fleming's plate could be produced only if the mould and the staphylococci were allowed to develop at rather a low temperature. Even room temperature in the summer would usually be too high, but here the vagaries of the English weather played their part. By examining the weather records at Kew, Hare discovered that for nine days after 28 July 1928 (just when Fleming had gone on holiday!), there was a spell of exceptionally cold weather. A final point is that the strain of penicillium which contaminated Fleming's plate is a very rare variety, and most penicillia do not produce penicillin in sufficient quantity to give rise to the

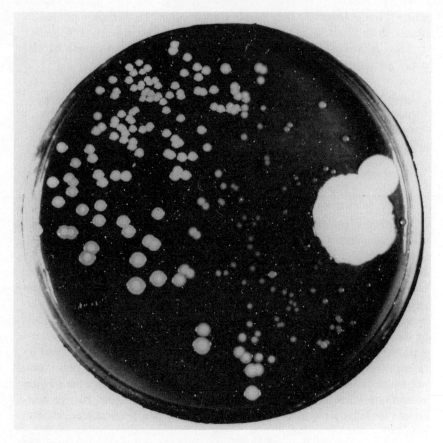

Plate 1 Fleming's penicillin culture-plate. Reproduced with the permission of St Mary's Hospital Medical School, London.

effect which Fleming observed. How did such a rare mould find its way into Fleming's laboratory? The most likely explanation is a curious one. There was at that time a theory that asthma was caused by moulds growing in the basements of the houses in which the asthmatics lived. This theory was being investigated by the scientist (C. J. La Touche) in the laboratory immediately below Fleming's, and La Touche had as a result a large collection of moulds taken from the houses of asthma sufferers. It seems probable that *penicillium notatum* was one of these moulds.

There is no doubt then that a great deal of luck was involved in the discovery of penicillin. Yet it still needed creativity and insight on Fleming's part to seize the opportunity which chance had presented to him. Nothing shows this more clearly than a comparison of Fleming's reaction to the contaminated plate with that of his colleagues in the laboratory (including the head of the laboratory, Sir Almroth Wright) when he showed it to them. With characteristic candour, Hare describes the complete lack of interest shown by himself and the others:

> The rest of us, being engaged in researches that seemed far more important than a contaminated culture plate, merely glanced at it, thought that it was no more than another wonder of nature that Fleming seemed to be forever unearthing, and promptly forgot all about it.
>
> The plate was also shown to Wright when he arrived in the afternoon. What he said, I do not recollect, but . . . one can assume that he was no more enthusiastic – he could not have been less – than the rest of us had been that morning. (1970, p. 55)

Fleming was by no means discouraged by his colleagues' cool reaction. He took a minute sample of the contaminating mould, and started cultivating it in a tube of liquid medium. At some later stage he photographed the plate, and made it permanent by exposing it to formalin vapour, which killed and fixed both the bacteria and the mould. Fleming kept the plate carefully, and it is now preserved in the British Museum. So we have here a case of fortune favouring the prepared mind. But what exactly had prepared Fleming's mind to realize something which his colleagues missed? We can answer this question by giving a brief account of the researches which had occupied Fleming in the fourteen years preceding his discovery of penicillin.

When the First World War broke out, Fleming was already working in the inoculation department headed by Sir Almroth Wright. Wright, Fleming, and the others were sent to Boulogne to deal with the war wounded, and, in particular, to try to discover the best way of treating infected wounds. At that time wounds were routinely filled with powerful antiseptics which were known to kill bacteria outside the body. Fleming, however, soon made the remarkable discovery that bacteria seemed to flourish in wounds treated with antiseptics even more than they did in untreated wounds. The explanation of this apparent paradox was quite simple. In an untreated wound the bacteria causing the infection were attacked by the body's natural defences, the white cells, or *phagocytes*, which ingested the invading bacteria. If the wound was treated with an antiseptic, some bacteria were indeed killed, but the protective phagocytes were also killed, so that the net effect was to make the situation worse than before. Wright and his group therefore maintained (quite correctly) that wounds should not be treated with antiseptics. They advocated the earliest possible surgical removal of all dead tissue, dirt, foreign bodies, and so forth, and then irrigating the wound with strong, sterile salt solution. The medical establishment of the day rejected this recommendation, and so the superior treatment was accorded only to those directly in the care of Wright and his team.

After the war, Fleming returned to the inoculation department in London, and here in 1921 he discovered an interesting substance which was given the name *lysozyme*. Lysozyme was capable of destroying a considerable range of bacteria, and was found to occur in a variety of tissues and natural secretions. Fleming first came across lysozyme while studying a plate-culture of some mucus which he took from his nose when he had a cold. He later discovered that lysozyme is to be found in tears, saliva, and sputum, as well as in mucus secretions. He extended his search quite widely in the animal and vegetable kingdoms, and found lysozyme in fish eggs, birds' eggs, flowers, plants, vegetables, and the tears of more than fifty species of animals. Lysozyme destroyed about 75 per cent of the 104 strains of airborne bacteria and some other bacteria as well. Moreover, Fleming was able to show that, unlike chemical antiseptics, even the strongest preparations of lysozyme had no adverse effects on living phagocytes, which continued their work of ingesting bacteria just as before. From all this, it seemed that lysozyme was part of many organisms' natural defence mechanisms against bacterial infection. Lysozyme had only one drawback. It did not destroy any of the bacteria responsible for the most serious infections and diseases. The hypothesis naturally suggested itself that the pathogenic bacteria were pathogenic partly because of their resistance to lysozyme.

If we put together Fleming's research on war wounds and his research on lysozyme, a problem situation emerges which I will call the 'antiseptic problem situation'. On the one hand, the chemical antiseptics killed pathogenic bacteria outside the body, but were less effective for infected wounds, partly because they destroyed the phagocytes as well. On the other hand, the naturally occurring antiseptic lysozyme did not kill the phagocytes, but also failed to destroy the most important pathogenic bacteria. The problem, then,

was to discover the 'perfect antiseptic' which would kill the pathogenic bacteria without affecting the phagocytes. The work on
lysozyme suggested that such an antiseptic might be produced by
some naturally occurring organisms.

It is commonly remarked that creativity consists in establishing a
hitherto unsuspected connection between two apparently different
areas, or problem situations. This was precisely what Fleming did
when he realized the significance of the penicillin plate. Instead of
dismissing the contaminated plate as a failure in his current investigation of the colours of staphylococci, he saw it as perhaps
providing the solution to the antiseptic problem situation which had
arisen from his earlier researches. In effect, he must have conjectured that the mould might be producing the 'perfect antiseptic'
capable of destroying pathogenic bacteria without disturbing the
phagocytes.

The assumption that Fleming made such a conjecture is borne
out by his subsequent actions. Fleming grew the mould on the
surface of a meat broth, and then filtered off the mould to produce what he called 'mould juice'. He then tested the effect of
this mould juice on a number of pathogenic bacteria. The results
were encouraging. The virulent streptococcus, staphylococcus,
pneumococcus, gonococcus, meningococcus, and diphtheria bacillus
were all powerfully inhibited. In fact, mould juice was a more
powerful germicide than carbolic acid. At the same time, mould
juice had no ill effects on phagocytes. Here at last seemed to be the
'perfect antiseptic'.

At this point in the story, however, a series of difficulties began
to emerge. Some further experimental work suggested, misleadingly, that penicillin might not be effective in the body. At the
same time, it proved difficult to isolate and purify the compound,
and there were difficulties about storing penicillin in such a way that
it would not rapidly lose its power to destroy bacteria. In his 1929
paper, Fleming wrote: 'It is suggested that it [penicillin] may be an
efficient antiseptic for application to, or injection into, areas infected
with penicillin-sensitive microbes' (p. 236). Yet the problems just
mentioned caused him to become despondent about the possible
uses of penicillin as an antiseptic; and not long after his paper was
completed, he abandoned research in that direction.

At this juncture, fortune once again favoured penicillin. Although
Fleming abandoned his earlier hopes that penicillin might be the
'perfect antiseptic', he found another practical use for it. This resulted
in a continued cultivation of the penicillin mould and production of

mould juice. Thus both mould and mould juice were readily available when Florey and his team at Oxford decided, a decade later, to
make another attempt to develop penicillin as an antiseptic.

The main source of income of the inoculation department where
Fleming worked was the production and sale of vaccines. There was
indeed an efficient unit for producing vaccine (a vaccine *laboratory*,
as it was then called) within the walls of the department, and
Fleming had been in charge of the production of vaccines since
1920. In particular, a vaccine was made against Pfeiffer's bacillus
(*Bacillus influenzae*) which was believed to cause influenza and
other respiratory infections. It was difficult to isolate this bacillus
because cultures were apt to be swamped by other micro-organisms.
Fleming, however, had discovered that penicillin, despite its effect
on so many virulent bacteria, left Pfeiffer's bacillus unaffected. By
incorporating penicillin into the medium on which he was growing
Pfeiffer's bacillus, he could eliminate the other germs, and produce
good samples of the bacillus itself. Fleming in fact used this method
for preparing the influenza vaccine in his vaccine laboratory, and
penicillin was made in the vaccine laboratory for this purpose every
week after its discovery. Significantly, the title of Fleming's 1929
paper on penicillin was: 'On the antibacterial action of cultures of a
penicillium with special reference to their use in the isolation of *B.
influenzae*'. Fleming also sent samples of the mould to other centres
concerned with the isolation of *B. influenzae*, and in this way cultures of the mould were established at the Lister Institute, Sheffield
University Medical School, and at George Dreyer's School of
Pathology in Oxford. Thus, when Florey and his team decided to
take up again the question of whether penicillin might be an efficient
antiseptic, they were able to find samples of Fleming's strain of
Penicillium notatum just down the corridor in the Dreyer School
of Pathology where they were working. Fleming's work, like
Kepler's, illustrates the close connection which so often exists between scientific discovery and practical applications.

Let us next try to examine the relevance of this important scientific
episode to the question of inductivism versus falsificationism. As a
matter of fact, Fleming's researches appear to involve *both* induction
and conjectures and refutations. On examining the contaminated
plate, Fleming arrived at two hypotheses. The weaker was that the
mould produced a bacteriolytic substance, while the stronger was
that this substance might be the 'perfect antiseptic' which his earlier
researches had led him to desire. Now here, observation definitely
preceded the formation of the hypotheses, and the hypotheses could

certainly not have been formulated without the observation. It thus seems correct to say that these hypotheses were obtained by induction from a chance observation. The hypotheses were, however, conjectures, and Fleming proceeded at once to test them by preparing his mould juice and investigating its effect on various kinds of pathogenic bacteria and on phagocytes. As we have seen, the early tests corroborated the 'perfect antiseptic' hypothesis, though some later results gave rather misleading counter-indications. In short, we seem to have a case of hypotheses generated by induction from observations and then developed by the method of conjectures and refutations. Does this mean that some kind of synthesis of inductivism and falsificationism is possible?

Such a synthesis has been suggested by Mitchell (1989).[3] Mitchell emphasizes what could be called *conjectural induction*, and describes some features of the process in the following two passages:

> It is a surprising characteristic of the conjectural process of induction that a useful theory appears to contain more information than the limited set of singular data from which it was originally induced. The useful theory must contain, not only the pattern of the singular data as presented, but also the pattern of some natural general principles of which the singular data were symptomatic. The task of the imaginative research worker is to guess what the general principles might be. (1989, p. 11)

And again:

> Although . . . the inductive process may proceed by small and apparently undramatic steps, these steps take the form of guesses, and the process of theory generation is non-deductive and basically conjectural in the sense described by Popper. (p. 12)

In fact, reflection shows that reconciling Popper's theory with some form of induction is not difficult. Popper gives no account of how scientific conjectures originate, and he considers this to be a matter for empirical psychology rather than philosophy of science. It is therefore possible to supplement his account by claiming that scientific conjectures sometimes originate by induction from observation. This would be Mitchell's conjectural induction. As we have seen, this view fits very well with the details of Fleming's discovery of penicillin. It is interesting that Mitchell speaks of 'the task of the *imaginative* research worker' (my emphasis); and indeed, I argued earlier that Fleming's conjectural induction involved considerable creativity. In cases like this, we could also use the term *creative induction*. We can, moreover, contrast it with another way of generating conjectures which could be called *creative theorizing*. In the latter case the conjecture is formed by meditating on earlier theoretical developments rather than by considering observations. A nice example of creative theorizing is provided by Copernicus. Copernicus was not much of an observational astronomer. His *De Revolutionibus Orbium Caelestium* of 1543 contains only twenty-seven observations made by himself, and neither these nor any other new observations seem to have had any influence in the genesis of his new hypothesis. Copernicus arrived at his heliocentric theory in something like the following way. He became dissatisfied with the Ptolemaic models used in the astronomy of his time. He therefore read many ancient Greek texts to see if an alternative approach could be found. In this way he lighted on the Pythagorean view that the Earth moved, and, starting with this hint, developed his own hypothesis. Here, then, we certainly have creative theorizing rather than any form of induction. Perhaps, however, Copernicus and Fleming occupy the two extreme points of a scale, and most hypothesis formation in science falls somewhere in between, being generated by reflection both on earlier theory *and* on new observations. Kepler in fact affords an example of such a mixture of creative theorizing and creative induction.

So far, then, all our examples have involved some human creativity; but the question naturally arises as to whether some form of *mechanical* or *Baconian induction* might not be possible as well. To pursue this question, we will examine in the next section another example – that of the discovery of the sulphonamide drugs.

2.7   The Discovery of the Sulphonamide Drugs: Mechanical or Baconian Induction

The sulphonamide drugs were discovered in Germany as a by-product of the activities of the giant chemical company I. G. Farben.[4] The discovery was made by a team headed by Gerhard Domagk, who was born in 1895 and appointed at the early age of thirty-two as director of research in experimental pathology and bacteriology in the institute attached to the I. G. Farben works at Elberfeld. Domagk and his team had huge laboratories in which they routinely tested compounds produced by the firm's industrial chemists on thousands of infected animals to see if the compounds had any therapeutic value.

The I. G. Farben chemists Hoerlien, Dressel, and Kothe produced a rich red dye which was very effective with protein materials such as wool and silk. This was known as *Prontosil rubrum*. Domagk and his team then discovered that this same compound possessed the definite ability to cure mice infected with haemolytic streptococci. Domagk published this finding in 1935, but referred back to experiments carried out in 1932.

This sequence of events can be described using the schema of conjectures and refutations. As each new compound was produced by I. G. Farben's industrial chemists, it was conjectured that it might have the ability to cure one or more bacterial infections. This conjecture was then tested by administering the compound to infected animals and seeing whether any improvement resulted. In the case of nearly all the compounds produced, the conjecture was refuted, but at last a compound appeared for which the corresponding conjecture was confirmed. In this example, then, the conjectures were not produced using any scientific imagination or creativity, but in a routine fashion. The process can therefore fairly be described as *mechanical*. But was it mechanical *induction*? The answer seems to be 'No'. At no point was there any inference of a conjecture from observations. The conjectures were generated in a routine fashion, and then tested experimentally. The observations came after the conjecture, as in Popper's model. I will, therefore, call this procedure *mechanical falsificationism* rather than *mechanical induction*. It should, however, be added that mechanical falsificationism bears some resemblance to the type of induction advocated by Bacon.

The discovery of the sulphonamide drugs does indeed have many Baconian features. To begin with, Bacon stresses the desirability of team work in the sciences. As he says:

> It is not a way over which only one man can pass at a time (as is the case with that of reasoning), but one in which the labours and industries of men (especially as regards the collecting of experience), may with the best effect be first distributed and then combined. For then only will men begin to know their strength when instead of great numbers doing all the same things, one shall take charge of one thing and another of another. (1620, p. 293)

Far from emphasizing imagination, creativity, or genius, Bacon seems to want to make science into a routine activity which could

be carried out by anyone of average intelligence. As he says: 'But
the course I propose for the discovery of sciences is such as leaves
but little to the acuteness and strength of wits, but places all wits
and understandings nearly on a level' (p. 270). Thus the army
of scientists in Domagk's laboratories performing routine tests on
routinely generated hypotheses seems to conform well to Bacon's
ideas. Moreover, all this activity did indeed result in what Bacon
called a 'secret of excellent use' (p. 292).

Bacon stressed that he wanted to introduce a new form of induction which was not induction by simple enumeration, but involved
exclusion and rejection. This is how he puts it:

> 'But the greatest change I introduce is in the form itself of induction
> and the judgment made thereby. For the induction of which the
> logicians speak, which proceeds by simple enumeration, is a puerile
> thing; concludes at hazard; is always liable to be upset by a con-
> tradictory instance; takes into account only what is known and
> ordinary; and leads to no result.
>
> Now what the sciences stand in need of is a form of induction
> which shall analyse experience and take it to pieces, and by a due
> process of exclusion and rejection lead to an inevitable conclusion. (p.
> 249)

By 'induction by simple enumeration' Bacon means something like
the induction from the observation of several thousand white swans
to the conclusion that the next observed swan will be white. Bacon
regards this as 'a puerile thing'. Now what happened in the case of
Prontosil rubrum was the 'exclusion and rejection' of a very large
number of compounds until eventually one of therapeutic value
was discovered. Because of all these parallels, I will use the term
Baconian induction as a synonym for *mechanical falsificationism*.
Admittedly this terminology could be a little misleading, since, as
already pointed out, Baconian induction is not really induction at
all. Bacon, moreover, fails to take account of a rather crucial difficulty which we must next consider.

Bacon seems to have thought that in any particular instance there
will be only a few possible hypotheses available. Thus, by quite a
short process of 'exclusion and rejection', we will be led to the truth
as 'an inevitable conclusion'. In fact, however, it is often possible to
generate with ease an enormous number of hypotheses, and there
may not be sufficient time or resources to test all of them in the
hope of hitting on one which works. Thus it would scarcely be
possible to test for therapeutic properties every single compound

which the chemists of today are capable of synthesizing. In this situation there has to be recourse to what are called *heuristics*.

A *heuristic* (from the Greek *heuriskein*, 'to discover') is a guide to discovery. In the context of *mechanical falsificationism*, hypotheses are generated by some routine or mechanical procedure; but in practice this procedure is unlikely to be totally random. It will usually be devised in accordance with some heuristic. Even the search procedure which lead to the discovery of *Prontosil rubrum*, the first of the sulphonamide drugs, was guided by various heuristics. One of these was the idea that dyes capable of staining textiles might also have useful therapeutic properties. The 'dye heuristic', as it might be called,[5] had been introduced before Domagk by Paul Ehrlich. Ehrlich discovered that if certain dyes are injected into living organisms, they are taken up and stain only some particular tissues and not others. Ehrlich gives the following example, which played an important role in his discovery of the method of 'vital staining':

> Thus, for example, methylene blue causes a really wonderful staining of the peripheral nervous system.
>
> If a small quantity of methylene blue is injected into a frog, and a small piece of the tongue is excised and examined, one sees the finest twigs of the nerves beautifully stained, a magnificent dark blue, against a colourless background. (1906, p. 235)

Ehrlich goes on to observe that this specific staining property is lost if the chemical composition of the dye is changed even to a small extent. Thus he says:

> I was able to prove that the nerve-staining property of methylene blue is conditioned by the presence of sulphur in the methylene-blue molecule. Synthetic chemistry has, in fact, given us a dye which, apart from the absence of sulphur, corresponds exactly in its chemical constitution to methylene blue. This is BINDSCHEDLER'S green. With the absence of the sulphur, there is associated the inability to stain living nerves. (Ibid.)

In the light of these interesting discoveries, Ehrlich reasoned somewhat as follows. Suppose we know that a particular disease is caused by the invasion of some micro-organisms. To cure the disease, we need to find a chemical which is highly toxic to these micro-organisms, but which does not harm the patient. This can be achieved if we can find a chemical which kills the micro-organisms *and* which is taken up only by the micro-organisms and not by the

other tissues. Now dyes like methylene blue are highly specific, in
that they are taken up by some tissues and not by others. Many dyes
are also toxic. So it is not unreasonable to think that some dyes
might have good therapeutic properties. Indeed, Ehrlich was able
to show that his favourite, methylene blue, was helpful in curing
malaria. As he says: 'In my further experiments . . . I started from
the supposition that dyes with a maximal tinctorial activity might
also have a special affinity for parasites within the host–organism, ...
I chose the malaria parasites and was able, in association with
Professor GUTTMANN, to show that methylene blue can cure
malaria' (p. 241).

So the 'dye heuristic' proved successful first for Ehrlich, then for
Domagk. Ironically, however, it turned out that the therapeutic
properties of *Prontosil rubrum* have nothing to do with its ability to
dye fabrics.

The molecular structure of *Prontosil rubrum* is shown in figure 2.2,
where the hexagons are the benzene rings whose discovery by
Kekulé we described earlier, and, as usual, N denotes one atom of
nitrogen, S of sulphur, O of oxygen, and H of hydrogen. It is clear
that the molecule consists of two halves joined by the double bond
denoted by $=$. In the body, four hydrogen atoms are added to the
molecule through the action of enzymes, and the molecule splits
into two different molecules, sulphanilamide and tri-amino-benzene
(see figure 2.3).

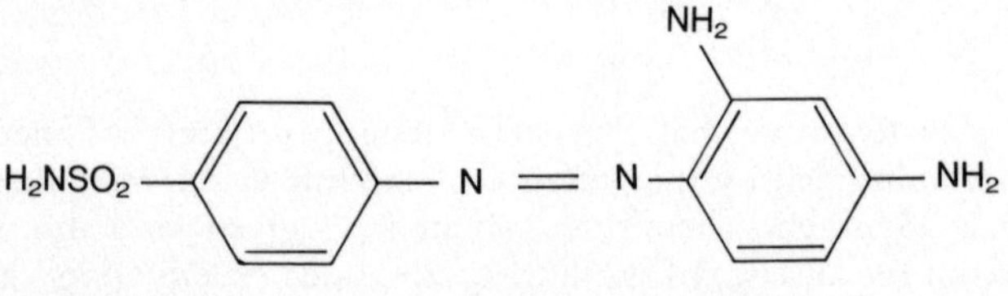

Figure 2.2   The molecular structure of *Prontosil rubrum*

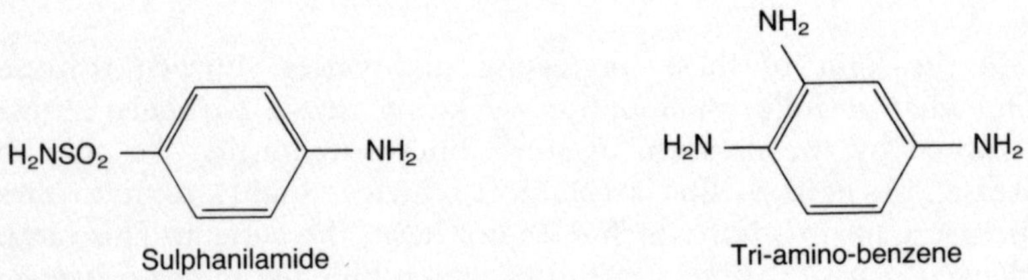

Figure 2.3   Reduction of *Prontosil rubrum* in the body

Now it turned out that only one of these molecules (sulphanil-amide) is responsible for killing the pathogenic bacteria. Sulphanil-amide, however, has no power of dyeing or staining either textiles or bacteria!

This concludes my account of Popper's critique of inductivism. In the next chapter I will consider Duhem's critique of inductivism, a critique which sheds light on some further aspects of the problem.

3

Duhem's Critique of Inductivism

3.1 Inductivism as the Newtonian Method

Duhem's critique of inductivism is contained in his masterly book on philosophy of science: *The Aim and Structure of Physical Theory*, which first appeared as a series of articles in the years 1904 and 1905. The attack on inductivism occurs in chapter 6, sections 4 and 5. Though only a few pages long, this is one of the most important passages in twentieth-century philosophy of science.

Duhem refers to inductivism as 'the Newtonian method', and he introduces it in the following way:

> It was this . . . that Newton had in mind when, in the 'General Scholium' which crowns his *Principia*, he rejected so vigorously as outside of natural philosophy any hypothesis that induction did not extract from experiment; when he asserted that in a sound physics every proposition should be drawn from phenomena and generalized by induction. (1904–5, pp. 190–1)

Newton's *Philosophiae Naturalis Principia Mathematica* ('Mathematical Principles of Natural Philosophy'), or *Principia* for short, was first printed in 1687. In it, Newton puts forward his three laws of motion and his law of gravity, and uses this system of theoretical mechanics to explain the movements of the solar system and to account for the tides and numerous other terrestrial phenomena. The great empirical success of Newton's theory led to its becoming accepted by virtually the entire scientific community in the early decades of the eighteenth century; and Newtonian mechanics remained the cornerstone of physics until the onset of the revolution in physics in the early years of the twentieth century.

Duhem is quite correct to link Newton with inductivism.

Newton includes an account of scientific method in his *Principia*, and this account is indeed inductivist in character. It is more doubtful whether Duhem is fair to call inductivism 'the Newtonian method'. As we have seen, the inductive method was formulated by Francis Bacon before Newton was born, and Bacon is indeed the probable source of Newton's views on method. Bacon had, in fact, studied at Trinity College, which was where Newton spent his Cambridge days.

Although inductivism should not be entirely identified with Newton's theory of scientific method, it will, none the less, be useful briefly to examine Newton's own version of inductivism, both because of its historical importance and because it is this version of inductivism which Duhem attacks. Newton's account of scientific method is set out in a section of the *Principia* which he entitles 'Rules of Reasoning in Philosophy'. Rule 3 is formulated as follows:

> *The qualities of bodies, which admit neither intensification nor remission of degrees, and which are found to belong to all bodies within the reach of our experiments, are to be esteemed the universal qualities of all bodies whatsoever.* (1687, p. 398)

This rule is clearly designed to infer universal laws or generalizations from finite sets of observations. Newton goes on to claim that he had obtained his law of gravity in this way. Newton's law of universal gravitation states that every body in the universe attracts every other body. The magnitude of this force of attraction between any two bodies is given by the product of their masses divided by the square of the distance between them. Thus if masses m_1 and m_2, are a distance r apart, the force of gravitational attraction between them is given by

$$F = Gm_1m_2/r^2,$$

where G is the universal constant of gravitation. Newton claims, in the following passage, to have inferred some aspects of this law inductively using his Rule 3.

> If it universally appears, by experiments and astronomical observations, that all bodies about the earth gravitate towards the earth, and that in proportion to the quantity of matter which they severally contain; that the moon likewise, according to the quantity of its matter, gravitates towards the earth; that, on the other hand, our sea

gravitates towards the moon; and all the planets one towards another;
and the comets in like manner towards the sun; we must, in consequence of this rule [i.e. Rule 3], universally allow that all bodies
whatsoever are endowed with a principle of mutual gravitation.
(p. 399)

In the next rule (Rule 4), Newton states his inductivism more
clearly and explicitly:

> In experimental philosophy we are to look upon propositions inferred by
> general induction from phenomena as accurately or very nearly true, notwith-
> standing any contrary hypotheses that may be imagined, till such time as
> other phenomena occur, by which they may either be made more accurate, or
> liable to exceptions.
>   This rule we must follow, that the argument of induction may not
> be evaded by hypotheses. (p. 400)

The aim of natural science (or 'experimental philosophy', as
Newton calls it) is to obtain propositions 'inferred by general induction from phenomena'.

Newton develops these ideas a little further in the General
Scholium to which Duhem refers. This was added to the *Principia* in
the second edition of 1713. Here, Newton says that although he has
discovered the laws governing gravitational attraction, he still does
not know the cause of gravity itself. As he puts it: 'Hitherto we
have explained the phenomena of the heavens and of our sea by the
power of gravity, but have not yet assigned the cause of this power'
(*Principia*, General Scholium, 1713, p. 546).

He then continues:

> But hitherto I have not been able to discover the cause of those
> properties of gravity from phenomena, and I frame no hypotheses;
> for whatever is not deduced from the phenomena is to be called
> an hypothesis; and hypotheses, whether metaphysical or physical,
> whether of occult qualities or mechanical, have no place in experi-
> mental philosophy. In this philosophy particular propositions are
> inferred from the phenomena, and afterwards rendered general by
> induction. Thus it was that . . . the laws of motion and of gravitation,
> were discovered. (p. 547)

Hypotheses then, for Newton, are speculations which are not
inferred by general induction from phenomena. He claims that he
himself does not put forward such hypotheses, and thinks that other
natural scientists should imitate his example.

It is worth making another couple of comments on this famous and interesting passage from Newton. First of all, Newton makes the typical Baconian inductivist's conflation of discovery and justification. The laws of motion and of gravitation were discovered by inductive inference from phenomena, and are also justified in the same way.

Secondly, at one point in the first sentence of the passage from page 547, Newton no longer speaks of inference by general induction from phenomena, but directly of deduction from phenomena. This is something of importance for, as Lakatos pointed out, induction and deduction are often conflated in the seventeenth and eighteenth centuries. As Lakatos himself says:

> In the seventeenth and eighteenth centuries there was no clear distinction between 'induction' and 'deduction'. (Indeed, for Descartes – *inter alios* – 'induction' and 'deduction' were synonymous terms; he did not think much of the relevance of Aristotelian syllogistic, and preferred inferences which increase logical content. Informal 'Cartesian' valid inferences – both in mathematics and science – increase content and can be characterized only by an infinity of valid patterns.) (1968, p. 130)

Lakatos perhaps exaggerates slightly here. After all, Hume, as we have seen, did have a concept of what could be obtained by reason, and denied that empirical laws or predictions could be derived in this way. Thus Hume, although he lived in the eighteenth century, does, in effect, distinguish between deduction and induction.

On the other hand, Lakatos is broadly correct. The major seventeenth-century thinkers (with the exception of Leibniz) regarded Aristotelian logic as just another piece of sterile scholasticism, and they no longer bothered to study it. A renewed interest in and development of formal logic did not occur until the time of Boole and Frege in the nineteenth century. In the seventeenth and eighteenth centuries, therefore, logic was treated informally, and such an informal approach does not bring out clearly the distinction between deduction and induction.

The situation was very different for Russell, his Cambridge followers, the Vienna Circle, and Popper. These twentieth-century thinkers had a very precise concept of deductive logic, as formalized, for example, in Russell and Whitehead's *PM*. They could thus make a very clear distinction between deductive and inductive logic. Perhaps the distinction was even a little too clear. This concludes our brief account of Newton's version of inductivism. We will next consider Duhem's criticisms.

3.2 Newton's Inference of the Law of Gravity from Kepler's Laws and Duhem's Objections

The phenomena from which Newton claimed to infer his law of gravity inductively included Kepler's laws. Indeed, Newton devotes a considerable section of the *Principia* to deriving the law of gravity from Kepler's laws. His argument is too technical for us to expound in detail, but we can explain the general idea in informal terms.

Let us begin with Newton's first law of motion, which he himself states as follows:

> *Every body continues in its state of rest, or of uniform motion in a right [i.e. straight] line, unless it is compelled to change that state by forces impressed upon it.* (1687, p. 13)

Now consider a planet, P, moving round the Sun, S. According to Kepler's laws, its orbit will be an ellipse with the Sun at one focus (see figure 3.1). Take the planet at a particular point in its orbit. If no forces were acting on it, it would, by Newton's first law of motion, continue moving with uniform velocity along the straight line shown as a series of dashes. However, it actually moves in an ellipse. So, Newton reasoned, a force must continually be acting on the planet, pulling it away from its natural straight line path, round into the curve. A mathematical calculation now shows that this force must be directed towards the Sun, and must vary inversely with the square of the distance of the planet from the Sun. Have we here valid inference of Newton's law of gravitation from the phenomena? Not according to Duhem, who writes:

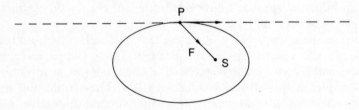

Figure 3.1   A planet moving round the Sun

Is this principle of universal gravitation merely a generalization of the two statements provided by Kepler's Laws and their extension to the motion of satellites? Can induction derive it from these two statements? Not at all. In fact, not only is it more general than these two statements and unlike them, but it contradicts them . . .

The principle of universal gravity, very far from being derivable by generalization and induction from the observational laws of Kepler, formally contradicts these laws. If Newton's theory is correct, Kepler's Laws are necessarily false. (1904–5, p. 193)

Duhem is quite correct in what he says here. If we consider the Sun and a planet in isolation from all the other bodies in the universe, then from Newton's theory it follows that the path of the planet will be an exact ellipse. However, it will be remembered that, according to Newton, every body in the universe attracts every other body. So our particular planet P will be attracted not just by the Sun, but by all the other planets of the solar system. These other gravitational attractions will perturb the orbit of the planet, causing its path to deviate slightly from a true ellipse. So, according to Kepler's laws, a planet moves round the Sun in an exact ellipse; whereas, according to Newton's theory, it moves on an ellipse with small perturbations due to the gravitational attractions of the other bodies in the solar system. Thus, Newton's theory formally contradicts Kepler's laws, and if Newton's theory is correct, Kepler's laws are necessarily false. It appears to follow from this that Newton's theory cannot be derived either inductively or deductively from Kepler's laws. If we suppose otherwise, we have an instance of deriving from given premises a conclusion that contradicts those premises; and this seems absurd.

Duhem goes on to observe that Newton's theory is actually validated by using it to calculate the perturbations of the planetary orbits, and then showing that these predicted perturbations agree with those actually observed. This is how he puts it:

Therefore, if the certainty of Newton's theory does not emanate from the certainty of Kepler's Laws, how will this theory prove its validity? It will calculate, with all the high degree of approximation that the constantly perfected methods of algebra involve, the perturbations which at each instant remove every heavenly body from the orbit assigned to it by Kepler's laws; then it will compare the calculated perturbations with the perturbations observed by means of the most precise instruments and the most scrupulous methods. (pp. 193–4)

When, later on, in 10.5, we come to examine the ways in which scientific theories can be confirmed by evidence, we will reach a conclusion which largely agrees with what Duhem says here.

Such, then, is Duhem's main argument against Newton's inductivism. He has, however, in the same place, a subsidiary argument which is itself of considerable interest.

Duhem argues that, in the course of his supposed derivation of the law of gravity, Newton has to re-describe Kepler's laws using the mechanical notions of 'force' and 'mass'. Yet this is a dubious move, since Kepler's laws can be stated in terms of positions, velocities, distances, areas, times, and so on, without ever mentioning forces and masses. As Duhem himself says: 'only dynamics permits us . . . to substitute statements relative to "forces" and "masses" for laws relative to orbits' (p. 194). The problem is how it is possible to derive, either inductively or deductively, a theory involving the new concepts of force and mass from a set of laws which do not involve these concepts. Any such thing seems indeed to be highly problematic.[1]

This concludes my account of Duhem's critique of inductivism. In the next section I will try to put this critique in its historical setting and to consider its validity, particularly in relation to the discussions of the previous chapter.

3.3   Criticisms of Inductivism and the Revolution in Physics

The enormous success of Newton's scientific work lent authority to the inductive method by which he claimed to have reached his results. It may therefore not be a coincidence that Duhem's and Popper's criticisms of inductivism are contemporary with the revolution in physics which showed that Newtonian mechanics was inadequate in some respects. I will next attempt to explore some possible connections.

The years 1904–6 saw the emergence of the special theory of relativity. Einstein published his account of the new theory in 1905. As we shall see in the next chapter, Poincaré shifted from defending Newtonian mechanics to arguing for its replacement between 1902 and 1904. Those who maintain that Poincaré was an independent discoverer of special relativity base their claim on his 1906 paper. Now, as we have seen, Duhem's *The Aim and Structure of Physical*

Theory first appeared as a series of articles in the years 1904–5, and was then published in book form in 1906. Is this just a temporal coincidence? Or was there some connection between Duhem's work and the beginning of the revolution in physics?

In fact, most of the ideas contained in *The Aim and Structure of Physical Theory* were introduced in a series of papers which Duhem published between 1892 and 1896, and which are conveniently collected in Jaki (ed.), 1987. However, as Brenner points out (1990b, pp. 330 and 334), these early articles do not contain Duhem's criticism of inductivism. This was first published in March and April of 1905. It is very unlikely that Duhem was influenced by Einstein, since, as we shall see later, Duhem continued as late as 1915 to reject Einstein's theory of relativity as an aberration of the German mind, and it is improbable that he read Einstein as early as 1905. On the other hand, it is very improbable that Duhem did *not* read some of Poincaré's reflections on the subject before 1905. Duhem undoubtedly followed Poincaré's work on physics and the philosophy of physics; and, indeed, many of Duhem's ideas emerged as developments or criticisms of Poincaré's views. In his 1904 lecture, Poincaré announced his conviction that Newtonian mechanics was inadequate to explain the new experimental findings in physics, and called for the development of a new mechanics. Perhaps it was reading this article which inspired Duhem to make his own critique not of Newtonian mechanics, but of the Newtonian method.

The first phase of the revolution in physics began with the emergence of the special theory of relativity around the year 1905. The second phase saw the appearance and development of quantum mechanics by Heisenberg, Schrödinger, Bohr, Dirac, and others in the years 1925–8. Once again, Newtonian mechanics was shown to be inadequate – this time in the micro-world of the atom. It seems not unreasonable to suppose that this second blow to Newtonian mechanics was one of the factors which stimulated Popper's critique of the Newtonian method (inductivism), which appeared in 1934. This conjecture agrees with Popper's own reminiscences, for Popper writes:

At the time (1930) when, encouraged by Herbert Feigl, I began writing my book, modern physics was in turmoil. Quantum mechanics had been created by Werner Heisenberg in 1925, but it was several years before outsiders – including professional physicists – realized that a major breakthrough had been achieved. And from the very beginning there was dissension and confusion. (1976, pp. 90–1)

Moreover, Popper's book *The Logic of Scientific Discovery*, published in 1934, contains not only his criticism of inductivism, but also a whole chapter (the ninth) devoted to a discussion of the philosophical problems of quantum theory.

Let us now try to evaluate Duhem's critique of inductivism, particularly with regard to the discussions of chapter 2. Inductivists such as Newton had high hopes of developing a method of induction which would be very similar to logical deduction. In logical deduction, the conclusion is derived from the premises, using a few simple and general rules. If the premises are accepted as certain, the conclusion too must be regarded as certain. The inductivist plan, then, was to develop a method of induction which would enable laws and theories to be derived from observations using a few simple and general rules. Even if these laws and theories could not be established with certainty, the hope was that they could be shown to hold, given the observations, with great probability. Newton argued that he had obtained his law of gravity from 'the phenomena', including Kepler's laws, by just such a method of induction. Duhem attacked this claim of Newton's. The heart of his criticism is contained in the following passage: '*The principle of universal gravity, very far from being derivable by generalization and induction from the observational laws of Kepler, formally contradicts these laws*' (1904–5, p. 193).

Duhem's point seems to me to tell heavily against the possibility of a method of induction analogous to logical deduction. It scarcely seems possible to derive, by anything like logical deduction, a conclusion which formally contradicts the premises. On the other hand, Duhem's criticism does not undermine the possibility of the kind of *conjectural induction* for which we argued in the previous chapter. There is no reason why a creative scientist should not from the study of a set of observations be led to a conjecture which partly contradicts the observations. Such a conjecture could be immediately tested by seeing whether the deviations from accepted observations which it predicted actually occurred. In fact, Newton seems to have obtained his law of gravity by a mixture of creative theorizing and conjectural induction. Roughly speaking, his creative theorizing from the earlier works of Descartes, Galileo, and Huygens led him to a system of mechanics which involved the new concepts of force and mass. Conjectural induction from Kepler's laws then led him to the conclusion that the force of gravity must vary as the inverse square of the distance. The simple account which I gave in 3.2 above probably corresponds to a part, though only a part, of his reasoning. A much fuller account of the historical origin of Newton's law of gravity is given by Duhem in 1904–5, pp. 220–52. Regarding this, Brenner shrewdly observes:

> It is by no means accidental that after rejecting the inductivist schema of the transition from Kepler's laws to Newton's principle in part two chapter six of *The Aim and Structure*, Duhem gives a long account of the historical genesis of the principle in the next chapter. This account is clearly intended as an alternative to the inductivist reconstruction. . . . History of science then provides the missing link for Duhem's rejection of inductivist methodology. (1990b, pp. 331–2)

As we might expect, Duhem's account on the whole emphasizes what we have called 'creative theorizing'; yet he sometimes mentions developments which could be described, in our terms, as 'conjectural induction'. Thus, for example, he says: 'Newton through his own efforts discovered the laws of uniform circular motion; he compared these laws . . . with Kepler's third law and recognized as a result of this comparison that the sun attracted equal masses of different planets with a force inversely proportional to the square of the distances' (1904–5, p. 251).

This concludes my exposition and evaluation of Duhem's critique of inductivism. It is just one of several important contributions by Duhem; we shall meet some of the others later in the book. In the next chapter we shall consider some of the contributions to philosophy of science of Duhem's brilliant contemporary Poincaré. It is therefore worth pausing now to say something about the lives of these men, who must surely be reckoned to be two of the most outstanding philosophers of science of the twentieth century.

3.4   The Lives of Duhem and Poincaré

Pierre Duhem was born in Paris on 10 June 1861, and died at his country home of Cabrespine (Aude) on 14 September 1916.[2] At the age of twenty, Duhem entered the École normale supérieure where he studied theoretical physics. He was a brilliant student, and received first place in 1885 in the competitive examination for teaching physics. Yet already Duhem had given offence to the French scientific establishment. The trouble began with a dissertation on thermodynamics composed when Duhem was twenty-three. This work contained very outspoken criticisms of Berthelot, then a powerful and influential figure, and it was, as a result, rejected. Moreover, no position in Paris became available for the young Duhem. After brief stays at the science faculties of Lille and Rennes, Duhem became professor of theoretical physics at Bordeaux at the age of thirty-two. He held this post till his death.

Though he later became reconciled with Berthelot, Duhem made further enemies, and was never on friendly terms with the Parisian scientific establishment. He was a man of strong personality, very honest, outstandingly brilliant intellectually, and with firmly held, but unusual, convictions. In short, he was just the sort of individual likely to come into conflict with an intellectual establishment. In addition to all this, Duhem was a devout Catholic, and held very conservative political views. He was therefore somewhat out of place in the liberal, anticlerical atmosphere of the Third Republic. Duhem is thus the mirror image of the members of the Vienna Circle, who were, for the most part, liberal and anticlerical, and consequently at odds with conservative and clerical circles in Austria.

Yet there is something to be said for Duhem's opponents, for Duhem was one of those unfortunate scientists who, despite great intellectual brilliance, seem to have an unfailing instinct for adopting approaches which prove to be unsuccessful. Duhem had a love of abstract mathematical theories, and tried to develop general thermodynamics and an energeticist programme similar to those of Ostwald and Mach. He rejected the attempt of Boltzmann and Gibbs to reduce thermodynamics to statistical mechanics, and attacked the introduction of atoms into physics. Yet, of course, it was the atomic approach which proved successful. Similarly, in the field of electricity, Duhem attacked Maxwell's electromagnetic theory, and supported the ideas of Helmholtz, which are now largely forgotten. He also failed to appreciate the importance of Lorentz's theory of electrons, and, as late as 1915, wrote a polemic against Einstein's theory of relativity. This is not to say that Duhem did not make some contributions to physics but, on the whole, he was unsuccessful in that field.

His studies of history and philosophy of science had a very different outcome. Duhem's first book on the history of science (*L'Évolution de la mécanique*) was published in 1903. This was followed by *Les Origines de la statique* (1905–6) and a massive, three-volume study of Leonardo da Vinci (1913). Also in 1913, there appeared the first volume of his monumental *Système du monde*. This was planned as twelve huge volumes covering the development of astronomy and physical theory from the pre-Socratics to Galileo.

By the time he died in 1916, Duhem had completed ten of the volumes single-handed, and had published five of these.

Duhem's main achievement in the history of science was a revaluation of the medieval period. Prior to him, science had been thought to have ended with the Greeks and to have begun again in the sixteenth century. Duhem showed that the medieval scholastics had considerable scientific achievements to their credit, and that these achievements had a notable influence on the scientific revolution of Copernicus and Galileo.

It is clear that Duhem's religious position influenced his work on the history of science. Enlightenment thinkers had seen Catholicism as an enemy of science, and had thought that science could flourish and develop only if it managed to shake itself free of the obstructive influence of the Church and its superstitious dogmas. Duhem wanted to show that, on the contrary, science had flowered in the medieval period under the aegis of the Church. He also wanted to show that the Enlightenment's hero Galileo had taken many of his ideas from the medieval scholastics, and that the Church had not been altogether wrong in its criticisms of Galileo's theories.

During this period of research on the history of science, Duhem was also developing his ideas on the philosophy of science. As we have seen, *The Aim and Structure of Physical Theory* appeared first as a series of articles in the *Revue de philosophie* in 1904 and 1905, then in book form in 1906, with a second edition in 1914.

Despite this intensive activity in history and philosophy of science, Duhem never abandoned physics, and indeed he always, rather perversely, regarded his work in physics as of much greater importance than his contributions to history and philosophy of science. Thus when, in the 1890s, an enquiry was made as to whether he would be interested in a professorship of the history of science at the Collège de France in Paris, he replied with the remark: 'I am a physicist. Paris will obtain me only as such, if I ever should return there.' Then again, when in 1913 he prepared his *Notice* in support of his candidacy for membership of the Academy, his account of his work in philosophy was only one tenth as long as his account of his work in physics, and was even shorter than his account of his work as a historian. On this point, the judgement of history is definitely against Duhem's assessment of himself.

It is interesting to compare Duhem with his great French contemporary Jules Henri Poincaré (1854–1912), for there exists a curious series of similarities and differences between the two thinkers. Poincaré, like Duhem, was trained in mathematics and physics, but, unlike Duhem, he was highly successful in these fields, and enriched them with a series of brilliant contributions. Indeed, he can lay claim, along with Einstein and Lorentz, to be one of the founders of the special theory of relativity. Duhem conforms quite closely to the romantic stereotype of the innovative thinker torn by inner contradictions and at odds with his social environment. Poincaré, by contrast, appears to have had a calm, well-balanced personality and to have fitted comfortably into his social milieu.

Poincaré was born at Nancy into what his obituarist in *The Times* describes as 'an old *bourgeois* family'. By contrast, Duhem's family was distinctly *petit bourgeois*, since his father was a textile salesman who never made much money. Poincaré began his education at the *lycée* of Nancy, and then passed with great distinction through the École polytechnique. He obtained his doctorate in 1879, and shortly afterwards obtained his first academic appointment at Caen. However, his stay in the provinces was a short one. Two years later, at the age of twenty-seven, he returned to Paris. From then on a succession of well-deserved honours were awarded to him. In 1887 he was elected a member of the Académie des sciences. In 1889 he won an international prize of 2,500 crowns and a gold medal offered by the King of Sweden for work on the three-body problem. The French government then made him a member of the Legion of Honour. In 1908 he was elected to the Académie française. Thus, while Duhem languished in the provinces, Poincaré established himself as the undisputed leader of French mathematics and physics in Paris.

Poincaré, like Duhem, began to take an interest in philosophical questions around 1900, and here, I think, we can see again the influence of the beginning of the twentieth-century revolution in physics. Between 1902 and his early death in 1912, Poincaré published three philosophical works: *Science and Hypothesis* (1902), *The Value of Science* (1905), and *Science and Method* (1908). After his death, his last thoughts on the subject were published in book form. Poincaré deals with philosophy of mathematics as well as philosophy of science; but, unlike Duhem, he never did research on the history of science.

Poincaré's philosophy of science is know as *conventionalism*, and I will describe it in chapter 4. It appeared at the same time and in the same social milieu as Duhem's, and there are, as we might expect, many similarities between their philosophical positions. Some writers have even classified Duhem's philosophy as a kind of conventionalism. This, however, seems to me a mistake.[3] Duhem often criticizes Poincaré in sharp terms: for example, 1904–5, p. 149f, where Duhem attacks an article of Poincaré's published in the *Revue de métaphysique et de morale* in 1902. Duhem's position could better be described as *modified falsificationism* than as conventionalism.

Let us move now from differences to similarities. Duhem and Poincaré both write with an admirable clarity and precision. Their elegance of style shows them to be the heirs, in this respect, of the classic French philosophical tradition of Descartes and Voltaire. Another characteristic common to Poincaré and Duhem is also one which distinguishes them from Russell and the Vienna Circle. This is the absence in their writings of any use of formal logic. Both men write very logically of course, but their logic is informal rather than formal. Duhem seems to have taken no interest in the new formal logic being developed by Frege, Peano, and Russell;[4] whereas Poincaré studied it, but rejected it as sterile and useless. Poincaré also rejected Russell's logicist view of mathematics, and conducted a controversy with Russell on this point. Poincaré's side of the debate is reprinted in his 1908 book *Science and Method* (part II, ch. 3–5). It constitutes a brilliant and witty polemic against Russell, Peano, and the other logisticians. For example, Poincaré writes: 'I find nothing in logistic for the discoverer, but shackles. . . . if it requires 27 equations to establish that 1 is a number, how many will it require to demonstrate a real theorem?' (p. 178). With characteristic insight, Poincaré comes close to anticipating Gödel's incompleteness theorem. Yet history has shown that his attacks on formal logic were misguided. The new logic is now constantly used throughout mathematics, and has become an indispensable tool for computer science.

Poincaré drew his inspiration for his philosophy of science and mathematics from the then current practice of science and mathematics in which he was actively engaged. Duhem, as we have seen in this chapter, used episodes from the history of science to criticize certain philosophical views and to support others. The contrast between Duhem and the Vienna Circle is an illustration of the difference between *historical* and *logical* approaches to the philosophy of science.

Those who adopt the logical approach attempt to give a logical analysis of scientific inference and of the structure of scientific theories. The title of Quine's book *From a Logical Point of View* (1953) clearly indicates the approach of its author. Carnap's *Logical Foundations of Probability* (1950) is an extreme example of the logical approach. This massive book is filled with formal logic and abstract symbolism, but there is no detailed discussion of even a single episode from the history of science.

The historical approach to philosophy of science is quite different in character. The adherents of this approach like to test their models of science by seeing how well they can account for case-studies in the history of science. If Carnap's *Logical Foundations of Probability* is the 'ideal type' of the logical approach, Kuhn's *The Structure of Scientific Revolutions* has a corresponding position in the historical approach. Kuhn actually begins his book: 'History, if viewed as a repository for more than anecdote or chronology, could produce a decisive transformation in the image of science by which we are now possessed' (1962, p. 1). He illustrates his philosophical views with a great mass of material from the history of science, but nowhere does he employ any of the symbolism or techniques of formal logic.

Much can be said for and against both these approaches. Those who favour the logical approach will no doubt accuse the historians of being vague and woolly, in contrast to their own precision and exactness. The historians might perhaps reply that the models of the logicians may be very precise, but they bear very little relation to any actual science, past or present. The ideal philosopher of science would no doubt combine elements from both approaches. The historical approach unquestionably produces greater realism, but the logical approach has received a boost from the ongoing computer revolution. Formal theories of science are tailor-made for implementation in the field of artificial intelligence, whereas the more intuitive insights of the historical school would need a great deal of processing before they could be used in this way. Perhaps the logical approach gives the science of computers, while the historical approach presents the science of human beings.

This is one example of how the development of artificial intelligence is affecting philosophy of science, but certainly not the only one. We have seen how inductivism was undermined by the revolution in physics. In the final section of this chapter, I will show how the contemporary revolution in computing is restoring an interest in inductivism.

3.5 Artificial Intelligence and the Revival of Inductivism

In recent years a branch of artificial intelligence has developed which is called *machine learning*. Those who study machine learning try to write programs which will enable a computer, when fed with data, to output a law or laws which these data satisfy. In other words, machine learning attempts to implement Baconian induction on a computer. Some workers in the field have even claimed remarkable successes. Thus Langley, Simon, Bradshaw, and Zytkow write: 'We will describe the computer program BACON.1, which as its name implies is a system capable of making scientific discoveries by induction on bodies of data' (1987, p. 25). Moreover, they go on to give the following list of physical laws discovered by BACON.1: Boyle's law, Kepler's third law, Galileo's law, and Ohm's law (p. 86). This makes it look as if Baconian induction has been definitely established as a reality. However, the claims made here (like so many in the field of artificial intelligence) must be treated with a certain degree of scepticism. After all, BACON.1 has only succeeded in discovering laws already known to those who wrote the computer program! Let us look at the case of Kepler's third law a little more closely, to see what has really been achieved.

Let D be the distance of a planet from the Sun, and P its period – that is to say, the time the planet takes to complete its orbit. Kepler's third law states that $D^3/P^2 = c$ (a constant) for all planets. The program BACON.1 is described in Langley *et al.*, 1987, pp. 66–86. In the case of Kepler's third law, the computer is given data consisting of values of D and P, and has the task of finding a law relating D and P. It is also given heuristics (pp. 66–7) which amount to asking it to look for a law of the form $D^m P^n = a$ constant, where m and n are integers. The program BACON.1 now enables the parameters m and n to be estimated from the data, and it comes up with $m = 3$, $n = -2$. Can we correctly describe this as remaking Kepler's discovery of his third law? It seems to me that we cannot.

The problem, of course, is that the computer is in effect told (i) which two variables to relate, and (ii) the general form of the law it should look for. The really difficult part of Kepler's discovery was finding the information in (i) and (ii), and this BACON.1 does not do at all. Once the information in (i) and (ii) is given, the problem reduces to estimating two parameters from the data. This step

BACON.1 does succeed in carrying out, but it is the easiest step; moreover, there is nothing surprising in a computer program estimating a few parameters in a given model.

To see more clearly the great difference between Kepler's original discovery and computer programs of the BACON type, it will be helpful to return briefly to our earlier example of Kepler's first law. We can then consider again the case of the third law. Kepler's first law states that all planets move in ellipses with the Sun at one focus. Kepler obtained this law by studying Tycho Brahe's data on the orbit of Mars. Let us next see why a normal scientist of Kepler's time, even if, *per impossibile*, he had had the help of a computer program much more powerful than BACON.1, would have failed to discover Kepler's first law.

Our hypothetical normal scientist would, of course, have adopted the dominant paradigm of his time – that is, Ptolemaic astronomy. Consequently he would have related the orbit of Mars to the Earth taken as fixed, and would have looked for a law consisting of a series of epicycles carried by a deferent. His hypothetical stronger BACON program could then have checked these deferent/epicycle models against the data by adjusting all the parameters – for example, radii of deferent and epicycles, their respective velocities, and so on – and have found the version which fitted best. No doubt, with a sufficient number of epicycles and powerful computing techniques, a good fit would have been found, but Kepler's first law would never have been discovered.

Kepler obtained his laws only because he had previously come to accept the new Copernican theory. The considerations which led him to do so were subtle and philosophical: a delicate, qualitative balancing of arguments for and against the new theory, an admiration for Pythagorean philosophy, and perhaps even Neoplatonic sun worship. All these heuristics, which were of vital importance historically, are absent from the BACON programs.

Moreover, even Copernicanism was not enough. To obtain his first law, Kepler had to take a further revolutionary step. He had to abandon the search for a law in terms of a combination of circular motions and try instead a law of a different form (the ellipse). This step contradicted two thousand years of astronomical tradition, and was not taken even by Copernicus or Galileo.

If we return to the case of Kepler's third law, we can see that the same points apply. The variables considered are D and P. But D, the distance of a planet from the Sun, would have had no significance in the dominant paradigm of Ptolemaic astronomy. No normal scientist of the time would have considered it. Still more remarkable was Kepler's search for a law of the form $x^m y^n = $ a constant. Ellipses had been studied in antiquity, but a law of the form $x^3 y^{-2} = $ a constant was a complete novelty. Of course, today, more than three and a half centuries after Kepler, laws of a generalized polynomial form have become a commonplace of scientific culture. This is apt to blind us to the greatness and originality of Kepler in introducing such a law for the first time.

The work of Langley, Simon, Bradshaw and Zytkow in their 1987 book is certainly of value and interest; but their claim to have produced computer programs capable of making major scientific discoveries by induction on bodies of data must be treated with some caution.[5] This is not to say that nothing has been achieved in the field of machine learning. On the contrary, Stephen Muggleton and his colleagues at the Turing Institute in Glasgow are developing what seems to be a much more hopeful approach to the subject.[6]

I will conclude the present chapter by considering whether our discussion of Baconian induction in 2.7 gives any indications about the possibility of implementing it on a computer. I argued earlier that *Baconian induction* is really equivalent to *mechanical falsificationism*, and can be considered as a procedure of conjectures and refutations. The conjectures are generated not by the creative insight of a scientist, but by some routine or mechanical procedure carried out in accordance with some *heuristic*. The conjectures formed in this way are then tested against the data until, hopefully, one of them turns out to fit the facts. How might such a procedure be implemented on a computer? The best approach would seem to involve co-operation between computer scientists and experts working in the field in question. The experts could provide the heuristics, and the computer scientists could then write programs for generating hypotheses in accordance with these heuristics and for testing these hypotheses against the data.

We can illustrate this approach by imagining a team of computer scientists transported back in time with their computer and an electricity generator to the year 1600 and given the task of helping the astronomers of that epoch to discover simple laws governing the motion of the planets. How would this team find the most appropriate heuristics for generating hypotheses? Well, they would clearly begin by interviewing the leading mainstream astronomers, and, as already pointed out, this would lead them to generating hypotheses in which the planets moved round a stationary Earth on paths which were a combination of circular motions. Some progress would undoubtedly be made along these lines, but the research programme would not lead to a major breakthrough. To obtain the breakthrough, the team would have to be a little more broad-minded and consult not just the establishment, but some of the outsiders dismissed by the majority as technically competent but crankish in their approach. Looking around among such individuals, the team would quickly come across Kepler, who with great enthusiasm would propose a quite different set of heuristics. He would suggest that, instead of relating all the variables to a stationary Earth, they should relate them to a stationary Sun. Instead of considering just circular motions, they should try other known curves such as the ellipses which had been studied in antiquity by Apollonius of Perga, and they might also try generating laws in the form of peculiar combinations of algebraic expressions. If the team were sensible enough to adopt the heuristics of this eccentric philosopher-mathematician, they would be quickly rewarded with success. It can be seen from this hypothetical example that there is no necessary antagonism between human creativity and computer approaches. On the contrary, human creativity is a wonderful resource and one which can be consciously used to produce better systems of artificial intelligence.

Part II
Conventionalism and the Duhem– Quine Thesis

4

Poincaré's Conventionalism of 1902

In this chapter I will give an exposition of Poincaré's philosophy of science, which is known as *conventionalism*. Poincaré was one of the leading mathematicians and physicists of his time, and in his mathematical work he made considerable use of non-Euclidean geometry. This led him to an interest in the nature and foundations of geometry. It is probable that he first devised conventionalism to give an account of geometry, and only later extended it to some other parts of science. As an introduction to Poincaré's ideas, therefore, it will be necessary to say a little about the philosophical influence of Euclidean geometry prior to the discovery of non-Euclidean geometry, and then about the impact of non-Euclidean geometry on the theory of knowledge. This is an important topic in itself, since Putnam has claimed that 'the overthrow of Euclidean geometry is the most important event in the history of science for the epistemologist' (1975, p. x).

Euclid's axiomatic-deductive development of geometry was written around 300BC, and was accepted, almost without challenge, as a true account of the geometry of space for more than 2,000 years. Moreover, Euclidean geometry was used by Newton in developing his mechanics, and so received additional support from the success of Newtonian physics. It is not surprising then that Euclidean geometry came to be seen as a certain and indubitable piece of knowledge. This attitude receives its classic expression in Kant's philosophy of geometry expounded in his *Critique of Pure Reason* (1781/7) and *Prolegomena* (1783). We will now briefly describe Kant's position.

4.1 Kant's Philosophy of Geometry

Kant's theory of geometry depends on a pair of distinctions, namely:

(i)   between *a priori* and *a posteriori* knowledge, and
(ii) between *analytic* and *synthetic* judgements.

The first distinction is really a traditional one. Kant explains it as follows: 'In what follows, therefore, we shall understand by *a priori* knowledge, not knowledge independent of this or that experience, but knowledge absolutely independent of all experience' (1781/7, A2/B3, p. 43). *A posteriori* knowledge, by contrast, does depend on experience.

The second distinction is really due to Kant himself – though there are traces of it in earlier authors. He says: 'Either the predicate B belongs to the subject A, as something which is (covertly) contained in this concept A; or B lies outside the concept A, although it does indeed stand in connection with it. In the one case I entitle the judgement analytic, in the other synthetic' (1781/7, A6/B10, p. 48).

Kant's notion of an analytic judgement can be illustrated by the favourite modern example (mentioned above): 'All bachelors are unmarried.' Here, the subject A is 'bachelors', and the predicate B is 'unmarried'. But bachelors are by definition unmarried men, so that the predicate is here (covertly) contained in the subject, and the judgement is therefore analytic.

Another approach, due to Frege, is to define an analytic judgement as one which can be reduced to a truth of logic using only explicit definitions. We have already mentioned the view held by Russell after 1900, and then by the Vienna Circle, that mathematics is reducible to logic. We now see that this view, known as *logicism*, can be formulated as the theory that mathematical propositions are analytic. This was *not* Kant's opinion regarding mathematics, which, for him, included Euclidean geometry as one of its most important parts. Kant believed that mathematical judgements were synthetic *a priori*.

Kant first argues that 'Properly mathematical propositions are always judgements *a priori*, and not empirical, because they carry with them necessity, which cannot be taken from experience' (1783, pp. 18–19). Kant thought that any generalisation based on experience such as 'All swans are white' cannot be necessarily true, since an exception, such as a black swan, might be found in the future. The theorems of Euclidean geometry, however, he regarded as necessarily true, and therefore as *a priori* rather than *a posteriori*. Consider, for example, the theorem that angles in a triangle are equal to 180°. This is not, on Kant's view, an empirical generalisation which might be contradicted by finding a triangle with angles adding up to 179°. The result is rather proved to be true by Euclid by showing that it follows logically from axioms which are themselves seen to be obviously correct. So the theorem is shown to be necessarily true, and since the necessity 'cannot be taken from experience', the judgement is 'a priori, and not empirical'.

Kant has next to show that the truths of Euclidean geometry are synthetic rather than analytic, which he tries to do with the following argument:

> Nor is any principle of pure geometry analytic. That the straight line between two points is the shortest is a synthetic proposition. For my concept of *straight* contains nothing of quantity but only a quality. The concept of the shortest is therefore wholly an addition, and cannot be drawn by any analysis from the concept of the straight line. (1783, p. 20)

These, then, are the considerations which convinced Kant that Euclidean geometry is synthetic *a priori*. But this doctrine is a puzzling one, and leads Kant immediately to raise the question of how such synthetic *a priori* knowledge is possible. This is indeed a problem. If a judgement is synthetic – that is, not based on a mere analysis of concepts – it would appear to be 'about the world', and hence knowable only on the basis of experience. How, then, can we know about the world *a priori*?

In his transcendental philosophy, Kant proposed some ingenious answers to this question, but my aim here is not to expound the Kantian system. I will therefore give next a brief account of how non-Euclidean geometry was discovered in the nineteenth century. We will then see why non-Euclidean geometry led Poincaré to reject Kant's view of geometry as synthetic *a priori* and give instead his own conventionalist account of geometry.

4.2   The Discovery of Non-Euclidean Geometry

Ironically enough, the discovery of non-Euclidean geometry arose out of attempts to make Euclidean geometry even more certain and well founded. Euclidean geometry was derived from five axioms, or postulates. Four of these did indeed seem to be obviously correct; but the fifth postulate, or parallel postulate, was a little less obvious than the others. Already some ancient Greek geometers had raised

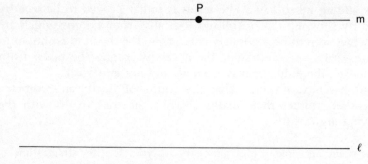

Figure 4.1   Parallel lines in Euclidean geometry

questions about this postulate, and the attempt was begun to improve on Euclid by deriving the fifth postulate from some even more evident axiom.

To see what was involved here, let us state the fifth postulate, not in the original form, but in a simpler, though equivalent form due to the British mathematician Playfair (1795). We will assume that we are dealing with plane geometry, so that all the points and lines considered lie in a plane. Playfair's form of the parallel postulate then becomes the following:

> Given a line l and a point P not lying on l, there exists one and only one line m through P parallel to l.

This is illustrated in figure 4.1. (Two lines are parallel if they do not meet, however far they are extended in either direction.)

If we think about the matter, there do appear to be two alternatives to this axiom. First of all, there might be no lines through P parallel to l. This alternative in fact gives rise to a form of non-Euclidean geometry known as Riemannian geometry. Secondly, there might be more than one line through P parallel to l. For example, in figure 4.2, m and m' might both be parallel to l. This alternative gives rise to a form of non-Euclidean geometry known as Bolyai–Lobachevsky geometry.

The first mathematicians to consider these possibilities, however, were very far from thinking that they could give rise to geometries alternative to Euclid's. On the contrary, they believed that these negations of the parallel axiom were absurd, and could quickly be shown to lead to a contradiction. Hence arose the plan of strengthening Euclid's parallel postulate using the method of *reductio ad absurdum*.

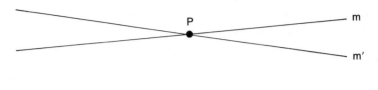

Figure 4.2   Parallel lines in Bolyai–Lobachevsky geometry

The two alternatives to Euclid's parallel postulate would be considered in turn, and would be shown to be absurd by deriving a contradiction from them. This would leave Euclid's parallel postulate as the only viable possibility.

This programme did have an initial success. It was shown that the assumption that there were no lines through P parallel to 1 did lead to a contradiction. In retrospect, this result was misleading, because it rested on the hidden premise that a straight line can be extended as far as we like in either direction. No one considered the possibility that there might be a finite upper bound to the length of any straight line; but this possibility, as we shall see, gives rise to Riemannian geometry.

At the time, however, the elimination of the first alternative to Euclid's parallel postulate seemed conclusive, and only the second alternative remained for consideration – namely, that through a point P not on 1 there might be more than one line parallel to 1. It proved much harder to derive a contradiction from this assumption. One of the most gallant attempts to do so was made by the Italian Jesuit Gerolamo Saccheri (1667–1733) in his book *Euclides ab Omni Naevo Vindicatus* ('Euclid Freed from Every Flaw') (Milan, 1733). Assuming the second alternative to Euclid's parallel postulate, Saccheri derived a succession of strange theorems, concluding eventually that these results were absurd and that Euclid must therefore be correct. But Saccheri's results, though strange, were not contradictory, and were indeed theorems of a non-Euclidean geometry. Yet such was the hold of Euclid on people's minds that almost a hundred years elapsed before any mathematician realised that this was the case.

The honour of publishing the first systems of non-Euclidean geometry is shared between a Russian, Lobachevsky, and a Hungarian, John Bolyai. They worked completely independently, and published within a few years of each other. Lobachevsky's 'Principles of Geometry' was printed in Russian in the *Kazan Bulletin* for 1829–30, while John Bolyai's 'The Science Absolute of Space' appeared as the appendix to a book by his father, Wolfgang Bolyai, on geometry, published in 1832. The great German mathematician Gauss had obtained similar results at an earlier date, but he did not publish because, as he said in a letter of 1829 to Bessel, he was afraid of the clamour of the Boeotians. Boeotia was a region of ancient Greece whose inhabitants were considered by the Athenians to be stupid and uncultured.

Like Saccheri, both John Bolyai and Lobachevsky took the second alternative to Euclid's parallel postulate and derived consequences from it, but their point of view was different. Saccheri had hoped to derive a contradiction from this alternative assumption, and hence to vindicate Euclid. Bolyai and Lobachevsky, however, regarded this assumption as one of the axioms of a geometry alternative to Euclid's. This is the form of non-Euclidean geometry now known as Bolyai–Lobachevsky geometry. In this geometry, the angles in a triangle are always less than 180° (or π radians). Indeed, the sum of the angles in a triangle is not constant (as in Euclidean geometry), but becomes smaller as the area of the triangle becomes bigger. To obtain the exact formula, we have to associate with each point a constant, K, known as the *curvature* at that point, and consider $\iint\limits_{A\ B\ C} K\, dS$ which is the integral of K taken over the area of the triangle ABC. We then have

$$\iint\limits_{A\ B\ C} K\, dS = A + B + C - \pi \tag{1}$$

In Bolyai–Lobachevsky geometry, K is always negative, so that $A + B + C < \pi$. In Euclidean geometry, $K = 0$, so that $A + B + C = \pi$ radians $= 180°$. The third case, in which K is always positive, gives Riemannian geometry, which we shall consider in a moment.

The publications of Bolyai and Lobachevsky did not in fact raise a clamour among the Boeotians. They were simply ignored. Bolyai did not publish anything further on the subject, whereas Lobachevsky, throughout the rest of his life, wrote a series of expositions and developments of his new geometry, including one in French, 'Géométrie imaginaire', published in Crelle's prestigious journal (1837), and one in German 'Geometrische Untersuchungen zur Theorie der Parallellinien', published as a booklet in Berlin (1840). Despite these efforts to overcome the language barrier, Lobachevsky did not succeed in arousing any interest among Western European mathematicians. This disappointed him considerably, but he continued undaunted and in the year before his death (1855), when already blind, he published in both French and Russian yet another exposition of his system – the Pangéométrie.

Non-Euclidean geometry did not become widely known until after the work of Riemann (1826–66), a student of Gauss and one of the most brilliant mathematicians of his time. Whereas Bolyai and Lobachevsky developed their system from a study of Euclid's parallel postulate, Riemann had a completely different approach. His ideas on non-Euclidean geometry originated by abstracting from some earlier work of Gauss on curved surfaces. This need not surprise us. As we have seen, work on the parallel postulate suggested, misleadingly, that a geometry in which there were no lines through P, a point not on l, parallel to l was impossible. Only a new approach showed that this was a possibility after all.

Riemann expounded his new system of non-Euclidean geometry in a famous lecture entitled 'Über die Hypothesen, welche der Geometrie zu Grunde liegen' (On the Hypotheses which lie at the Foundations of Geometry), delivered as qualifying lecture for the title of *Privatdozent* to the faculty at Göttingen in 1854. Gauss himself was present at the lecture.

In Riemannian geometry, there are no parallel lines, so all straight lines intersect each other. Straight lines are not infinitely long, and, indeed, there is a finite upper bound on the length of straight lines. In Euclidean geometry, there is one and only one straight line between any two distinct points A and B. In Riemannian geometry, there may be more than one straight line between two points, so that two straight lines can enclose a space. This last property is impossible to represent visually on a plane sheet of paper. If we draw two 'straight' lines l, m, between the points A and B (see figure 4.3), the lines look curved, not straight. The kind of difficulty seen here should perhaps give the reader some sympathy for those who rejected non-Euclidean geometry. Finally, in Riemann geometry, the angles in a triangle are always greater than $180°$ (or π radians). The formula is again (1) except that, in Riemannian geometry, K is always positive, so $A + B + C > \pi$. The bigger the triangle, the greater is the difference between the sum of its angles and $180°$.

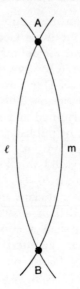

Figure 4.3   Two lines enclosing a space

Riemann was in a much better position than Bolyai or Lobachevsky to get his ideas known to the mathematical community. Germany was the leading country in the world at that time for mathematical research. Riemann was a pupil of the most famous German mathematician, Gauss, and was working at the leading German university for mathematical research, Göttingen. None the less, more than a decade was to pass after his lecture of 1854 before the ideas of non-Euclidean geometry became widely known among mathematicians.

One reason for this delay was that Riemann's key lecture of 1854 was published only in 1867, one year after its author's early death. Generally, however, in the 1860s, the tide at last began to turn in favour of non-Euclidean geometry. In the years 1860 to 1863, the correspondence between Gauss and a certain Schumacher was published. The frequent references to the unknown names of Bolyai and Lobachevsky aroused interest, and caused Hoüel to undertake the French translation of the works of these men. In 1866 there appeared the French translation of Lobachevsky's 1840 German booklet, together with extracts from the Gauss—Schumacher correspondence; while in 1867 the French translation of Bolyai's paper was published. Finally, in the years 1868–72, the Italian mathematician Beltrami and the German mathematician Klein published proofs of the consistency of the various forms of non-Euclidean geometry relative to Euclidean geometry. The old hope of Saccheri and others that a contradiction would be found in non-Euclidean geometry was shown to be an illusion, and, from a logical point of view, non-Euclidean geometries were proved to be on a par with Euclidean geometry.

Many of the relative consistency proofs of Beltrami and Klein are mathematically quite complicated, but there is one particularly simple such proof which I will give in outline. It shows the general method of these relative consistency proofs, and illustrates other points connected with non-Euclidean geometry as well. This proof applies to a particular kind of Riemannian geometry known as *double elliptic geometry*.

Let us suppose that we have an axiomatic system for double elliptic Riemannian geometry in the plane. The idea of the relative consistency proof (in this case, as in others) is to produce a model of this system within Euclidean geometry. This model consists of a kind of dictionary according to which the terms of double elliptic Riemannian geometry, such as point, line, and so on, are translated into Euclidean geometry. This translation has to be of such a kind that all the axioms of Riemannian geometry are true in Euclidean geometry when translated according to the dictionary.

In this simple case, the model for the whole plane of the Riemannian geometry is the surface of a sphere in three-dimensional Euclidean space. A point in the Riemannian plane corresponds to a point on the sphere. A straight line in the Riemannian plane corresponds to a great circle on the sphere. (A great circle is a circle on the sphere whose centre is the centre of the sphere.) The angle between two straight lines intersecting at a point is simply the angle between two intersecting great circles. This is illustrated (in figure 4.4) by the diagram of a sphere which can be thought of as the Earth. Two great circles have been drawn through the North and South poles (N and S) which intersect at right angles. These intersect another great circle (the Equator) at A and B, and the angles NAB and NBA are also right angles.

The axioms of double elliptic Riemannian geometry are all satisfied in this model. Without giving a full proof, we can make this plausible by considering the apparently paradoxical features of Riemannian geometry listed earlier. First of all, it is true in the model that there are no parallel lines, since any two great circles intersect. Moreover, there is clearly a finite upper bound to the

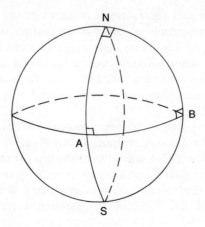

Figure 4.4   A model for double elliptic geometry

length of any great circle. More than one – indeed, an infinite number – of great circles pass through the points N and S, while the great circles NAS and NBS clearly enclose a space. Finally, in the triangle NAB, all the angles are right angles, so that the sum of the angles in the triangle is 90° + 90° + 90° = 270°, which is clearly greater than 180°.

The curvature $K > O$ of the Riemannian geometry becomes the curvature of the sphere. This curvature is the same at every point on the sphere, so we are dealing here with a space of constant curvature. There are also Riemannian spaces of variable curvature, in which the curvature K varies from point to point.

We can use this model to show that if Euclidean geometry is consistent, then so is double elliptic Riemannian geometry. For suppose, as Saccheri would have hoped, that we can derive a contradiction from the axioms of this Riemannian geometry. Take this proof of the contradiction and translate it into the model, using our dictionary. We then obtain a proof of a contradiction using only assumptions which are true in Euclidean geometry, and thus Euclidean geometry is also shown to be contradictory.

This model has a certain natural feel. Consider beings who are confined to the surface of a sphere. We ourselves are for the most part in this condition. For such beings, the shortest practical path between two points A and B on the sphere would be a great circle connecting A and B. The Euclidean straight line would be a 'tunnel' between A and B, whose construction we may suppose to be hardly practical. If these beings defined a straight line as the shortest path on the sphere between two points, they would have Riemannian rather than Euclidean geometry. The model therefore gives Riemannian geometry a homely feel. It is harder, however, to imagine three-dimensional space being Riemannian rather than Euclidean. To do so, we have to suppose that three-dimensional space is curved in the fourth dimension. This is impossible to visualize, but can be described mathematically by means of equations. This leads us to the conception of the geometry of space which underlies Einstein's general theory of relativity.

In 1915 Einstein introduced his general theory of relativity, in which the geometry of physical space is assumed to be Riemannian rather than Euclidean. The new theory explained the motion of the perihelion of Mercury, which was an anomaly for Newtonian theory. (I will give more details about this in the next chapter.) Four years later, the new theory received further confirmation from the eclipse experiment in which the predictions of general relativity were shown to be much more accurate than those of Newtonian theory. Thus, from the 1920s, general relativity came to be accepted as more accurate than Newtonian theory, thereby showing that the true geometry of physical space was non-Euclidean rather than Euclidean.

It is important to note for what follows that the Riemannian geometry used by Einstein was one of variable, rather than constant, curvature. General relativity postulates an interaction between matter and space, such that space becomes curved near very large gravitating masses like the Sun. For short distances in relatively weak gravitational fields, however, Euclidean geometry remains approximately true.

This concludes our account of the discovery and eventual triumph of non-Euclidean geometry.

4.3 Poincaré's Conventionalist Philosophy of Geometry

We how have enough background to understand the conventionalist account of geometry which Poincaré expounded in his 1902 book *Science and Hypothesis*. As with any philosophical text, it is important to bear in mind the exact historical conjuncture at which it was written and the corresponding background knowledge assumed by its author. As we have seen, non-Euclidean geometries had been proved consistent relative to Euclidean geometry by the early 1870s. A leading mathematician such as Poincaré, who had worked extensively with non-Euclidean geometries, would in 1902 have taken their logical possibility completely for granted. On the other hand, 1902 is thirteen years before the appearance in 1915 of the general theory of relativity, with its postulate that Riemannian geometry is the true geometry of space. Indeed, in 1902 nearly all mathematical physicists, including Poincaré, would have regarded non-Euclidean geometries as logically possible but not physically real. It was still almost universally accepted that the true geometry of space was Euclidean. We shall see in a moment how this assumption affected Poincaré's reflections on the nature of geometry.

Poincaré regards non-Euclidean geometry as refuting the Kantian position. He writes: 'what is the nature of geometrical axioms? Are they synthetic *a priori* intuitions, as Kant affirmed? They would then be imposed upon us with such a force that we could not conceive of the contrary proposition, nor could we build upon it a theoretical edifice. There would be no non-Euclidean geometry' (1902, p. 48).

However, Poincaré is just as much opposed to empiricism as he is to Kantianism. He goes on to say: 'Ought we, then, to conclude that the axioms of geometry are experimental truths? But we do not make experiments on ideal lines or ideal circles; we can only make them on material objects'(p. 49).

There are various replies which empiricists could make to this argument. They could say that geometry is an empirical theory of actual space, but that, like many theories of physics – for example, the ideal gas laws – it must be understood as holding only approximately. Alternatively, they could accept the physical reality of the spatial continuum as described by geometry, but argue that our evidence for the exact truth of geometry consists of inexact observations – for example, of material objects which are approximate realizations of straight lines, circles, and so forth. Such observations only partially confirm the truth of geometry (interpreted as holding exactly) – but then no scientific theory is ever confirmed more than partially.

It is interesting to compare Poincaré's total rejection of Kantianism and empiricism regarding geometry with the position adopted by Russell only five years earlier in his *Essay on the Foundations of Geometry* of 1897.

Russell, like Poincaré, thought that Kant's view of Euclidean geometry as *a priori* is untenable in the light of non-Euclidean geometry. However, unlike Poincaré, he believed that something of

Kant's position on geometry could be rescued. More specifically, he used an argument along Kantian lines to try to prove the *a priori* validity of a weaker sub-system of Euclidean geometry: namely, the sub-system which deals with the relations of points, lines, and planes, but *without* introducing the metrical notions of distance and angle. This sub-system is known as *projective geometry*. So Russell's claim is that projective geometry rather than Euclidean geometry is *a priori*. In chapter 3, section A, of his book, he attempts to give a proof of this, and concludes as follows:

> With this, our deduction of projective Geometry from the *a priori* conceptual properties of a form of externality is completed. . . . I wish to point out that projective Geometry is wholly *a priori*; that it deals with an object whose properties are logically deduced from its definition, not empirically discovered from data; that its definition, again, is founded on the possibility of experiencing diversity in relation, or multiplicity in unity; and that our whole science, therefore, is logically implied in, and deducible from, the possibility of such experience. (1897, p. 146)

Within projective geometry, we can go on to give definitions of the metrical notions of distance and angle; but these definitions can be formulated in various ways. One particular set of definitions yields ordinary Euclidean geometry, while others give various systems of non-Euclidean geometry. According to Russell, we cannot decide *a priori*, but only on empirical grounds, which of these various geometries is true of actual space. He writes: 'The Euclidean and non-Euclidean spaces give the various results which are *a priori* possible; the axioms peculiar to Euclid – which are properly not axioms, but empirical results of measurement – determine, within the errors of observation, which of these *a priori* possibilities is realised in our actual space' (p. 147). Russell here expresses his belief that the empirical results of measurement support the claim that actual space is Euclidean.

So, while Poincaré rejected both Kantianism and empiricism, Russell devised a view which combined elements of both positions. Russell used an argument along Kantian lines to try to establish that projective (*not* Euclidean) geometry is *a priori* correct. This left open a range of possibilities, both Euclidean and non-Euclidean, concerning the geometry of actual, physical space; and Russell held that the choice between these various possibilities could be made only on empirical grounds by observation and experiment.

Russell's view was certainly very reasonable when he propounded it, but it did not fare well in the light of Einstein's general theory of relativity. The problem was that, as we have already mentioned, the general theory of relativity involves Riemannian geometry of *variable* curvature; whereas Russell's doctrine that projective geometry was *a priori* correct ruled out as *a priori* impossible Riemannian geometry of variable curvature and allowed only Riemannian geometry of *constant* curvature. So although Russell's theory allows several different types of non-Euclidean geometry, it actually rules out the particular one which was used by Einstein.

In his *Essay on the Foundations of Geometry*, Russell quite explicitly criticised Riemann for allowing spaces of non-constant curvature. In chapter 3, section B, Russell gives an alleged proof of the *a priori* truth of the axiom of free mobility – that is, the axiom which postulates that spatial magnitudes can be moved from place to place without distortion. From this axiom it follows that spaces of non-constant curvature are impossible. Russell accordingly says:

> Riemann has failed to observe, what I have endeavoured to prove in the next chapter, that, unless space had a strictly constant measure of curvature, Geometry would become impossible; also that the absence of constant measure of curvature involves absolute position, which is an absurdity. Hence he is led to the conclusion that all geometrical axioms are empirical, and may not hold in the infinitesimal where observation is impossible. (1897, p. 69)

Russell was never dogmatic in his attitudes, and he welcomed the general theory of relativity with enthusiasm, repudiating his earlier views on geometry. In any case, by that time Russell had eliminated all traces of Kantianism from his philosophy of mathematics, and was the leading advocate of the *logicist* view of mathematics.

Let us, however, return to Poincaré. Having rejected empiricism and Kantianism, he proceeded to expound his own conventionalist philosophy of geometry:

> *The geometrical axioms are therefore neither synthetic a priori intuitions nor experimental facts.* They are conventions. Our choice among all possible conventions is *guided* by experimental facts; but it remains *free*, and is only limited by the necessity of avoiding every contradiction, and thus it is that postulates may remain rigorously true even when the experimental laws which have determined their adoption are only approximate. In other words, *the axioms of geometry* (I do not speak of those of arithmetic) *are only definitions in disguise.* What, then, are we to think of the question: Is Euclidean geometry true? It has no

meaning. We might as well ask if the metric system is true, and if the old weights and measures are false; if Cartesian co-ordinates are true and polar co-ordinates false. One geometry cannot be more true than another; it can only be more convenient. (1902, p. 50)

Poincaré's conventionalist account of geometry has two parts. First there is the claim, which we have just quoted, that Euclidean geometry is a set of conventions rather like those which define the metric system. Second, however, Poincaré holds that Euclidean geometry is the simplest possible convention and that, since it agrees well enough with observation, it will never be given up. Poincaré believes that the simplicity of Euclidean geometry is an objective mathematical fact. What he probably has in mind is that in Euclidean geometry the curvature of space is zero ($K = 0$), which might arguably be considered as objectively simpler than K having any non-zero value. At any rate, this is what he says:

> Now, Euclidean geometry is, and will remain, the most convenient: 1st, because it is the simplest, and it is not so only because of our mental habits or because of the kind of direct intuition that we have of Euclidean space; it is the simplest in itself, just as a poiynomial of the first degree is simpler than a polynomial of the second degree; 2nd, because it sufficiently agrees with the properties of natural solids, those bodies which we can compare and measure by means of our senses. (p. 50)

Poincaré buttresses this later on by appealing to a type of argument introduced by his contemporary Duhem. We shall consider the Duhem thesis in the next chapter, but Poincaré's use of it forms a convenient introduction to what will be said there. Poincaré writes:

> If Lobatschewsky's geometry is true, the parallax of a very distant star will be finite. If Riemann's is true, it will be negative. These are the results which seem within the reach of experiment, and it is hoped that astronomical observations may enable us to decide between the two geometries. But what we call a straight line in astronomy is simply the path of a ray of light. If, therefore, we were to discover negative parallaxes, or to prove that all parallaxes are higher than a certain limit, we should have a choice between two conclusions: we could give up Euclidean geometry, or modify the laws of optics, and suppose that light is not rigorously propagated in a straight line. It is needless to add that every one would look upon this solution as the more advantageous. Euclidean geometry, therefore, has nothing to fear from fresh experiments. (pp. 72–3)

Poincaré is of course correct in saying that to test Euclidean geometry experimentally, we have to assume more than just the axioms of Euclidean geometry. We need auxiliary assumptions such as that the path of a light ray is a straight line. It is thus always logically possible, in the face of an apparent experimental refutation, to give up an auxiliary assumption rather than Euclidean geometry itself. However, Poincaré was wrong in thinking in 1902 that the scientific community would hold on to Euclidean geometry, come what may. As a matter of fact, the community decided, after the introduction of the general theory of relativity in 1915 and its experimental confirmation in the subsequent few years, to limit the applicability of Euclidean geometry and to regard Riemannian geometry with non-constant curvature as giving a more accurate account of the geometry of physical space.

This, however, is to move our historical point of view to a date after 1920 and to judge the matter with historical hindsight. Let us now return to the early years of the twentieth century and continue our account of Poincaré's views in 1902 by explaining how he extended his conventionalist philosophy from Euclidean geometry to Newtonian mechanics.

4.4 Poincaré's Conventionalism and Newtonian Mechanics

Poincaré argues that the laws of Newtonian mechanics, like the axioms of Euclidean geometry, are conventions. His general line of argument is exactly the same for Newtonian mechanics as it was for Euclidean geometry. He begins by claiming that the laws of mechanics are neither *a priori* nor experimental truths. Although these laws were suggested by experience, future experiments and observations can never invalidate them. The laws of Newtonian mechanics are disguised definitions, or conventions, and, moreover, the simplest such conventions.

Poincaré considers the laws of Newtonian mechanics in turn. Let us follow his argument in more detail in the case of Newton's first law of motion, or the principle of inertia, which we discussed in chapter 3.

Poincaré begins by arguing that the principle of inertia is not an *a priori* truth. He says:

The Principle of Inertia – A body under the action of no force can only move uniformly in a straight line. Is this a truth imposed on the mind *à priori*? If this be so, how is it that the Greeks ignored it? How could they have believed that motion ceases with the cause of motion? or, again, that every body, if there is nothing to prevent it, will move in a circle, the noblest of all forms of motion? (1902, p. 91)

Poincaré's statement of the beliefs of the Greeks is perhaps a little misleading. According to Aristotle, only bodies in the heavenly region, and hence composed of the fifth heavenly element (aither), moved naturally in a circle. Poincaré's general argument is, however, convincing. Aristotelian mechanics is quite different from Newtonian mechanics, and yet Aristotelian mechanics was believed to be correct for many centuries. It is hard to see how this historical fact is compatible with Newton's laws of motion being *a priori* truths.

Having denied that the principle of inertia is an *a priori* truth, Poincaré goes on immediately to deny that it is an experimental fact:

Is, then, the principle of inertia, which is not an *à priori* truth, an experimental fact? Have there ever been experiments on bodies acted on by no forces? and, if so, how did we know that no forces were acting? The usual instance is that of a ball rolling for a very long time on a marble table; but why do we say it is under the action of no force? Is it because it is too remote from all other bodies to experience any sensible action? It is not further from the earth than if it were thrown freely into the air; and we all know that in that case it would be subject to the attraction of the earth. (pp. 91–2)

Moreover, Poincaré argues that if the principle of inertia were an experimental law, it might, in future, be modified in the light of observation and experiment and replaced by a more accurate law. But Poincaré thinks that the revision of the laws of Newtonian mechanics, like the revision of the axioms of Euclidean geometry, is not a serious possibility. As he says: 'An experimental law is always subject to revision; we may always expect to see it replaced by some other and more exact law. But no one seriously thinks that the law of which we speak will ever be abandoned or amended. Why? Precisely because it will never be submitted to a decisive test' (pp. 95–6). Suppose, for example, we observe what seems to be a deviation from the principle of inertia. Such an apparent deviation, Poincaré argues, need never force us to abandon the principle of inertia, because we can always get round the difficulty by postulating that the deviation is due to invisible molecules:

> If, then, the acceleration of bodies we cannot see depends on something else than the positions or velocities of other visible bodies or of invisible molecules, the existence of which we have been led previously to admit, there is nothing to prevent us from supposing that this something else is the position or velocity of other molecules of which we have not so far suspected the existence. The law will be safeguarded. (p. 96)

Poincaré then considers the other laws of Newtonian mechanics, and reaches the conclusion that: 'The principles of dynamics appeared to us first as experimental truths, but we have been compelled to use them as definitions' (p. 104), and again that: '*Experiment may serve as a basis for the principles of mechanics, and yet will never invalidate them*' (p. 105). Poincaré summarizes his overall view regarding the principles of Newtonian mechanics as follows:

> If these postulates possess a generality and a certainty which was absent in the experimental truths from which they were derived, it is because they reduce in final analysis to a simple convention that we have a right to make, because we are certain beforehand that no experiment will come to contradict it. This convention, however, is not absolutely arbitrary; it is not the child of our caprice. We admit it because certain experiments have shown us that it will be convenient, and thus is explained how experiment has been able to build up the principles of mechanics, and why, nevertheless, it cannot overthrow them. (p. 136; I have here altered the standard English translation slightly to give a more accurate rendering of the French.)

Poincaré's conventionalist account of Newtonian mechanics, like his conventionalist account of Euclidean geometry, is no longer tenable in the light of the twentieth-century revolution in physics. The view that Newtonian mechanics holds exactly in all circumstances was, contrary to Poincaré's predictions, abandoned because of the development of relativistic mechanics. Indeed, Newtonian mechanics is now thought to hold only approximately for bodies moving at speeds much less than the speed of light and in relatively weak gravitational fields. Newtonian mechanics was not in fact saved by the postulation of invisible molecules or any similar device.

But here we come to a paradox, for Poincaré, in his work as a physicist, was one of the principal initiators of the twentieth-century revolution in physics. What happened was that Poincaré changed his mind regarding mechanics between 1902 and 1904. In the introduction to his 1905 book *The Value of Science*, Poincaré explicitly draws attention to this change of opinion. He writes:

> It was altogether natural . . . that celestial mechanics should be the first model of mathematical physics; but . . . this science . . . is still developing, even rapidly developing. And it is already necessary to modify in certain points the scheme I outlined in 1900 and from which I drew two chapters of 'Science and Hypothesis.' In an address at the St Louis exposition in 1904, I sought to survey the road travelled; the result of this investigation the reader shall see farther on. (1905, pp. 13–14)

In fact, Poincaré's St Louis address of 1904 is reprinted as chapters 7–9 of his 1905 book. It is in this address that we find the change in his opinion regarding Newtonian mechanics. Poincaré states quite clearly that this change of opinion was brought about by reflection on some new experimental results. As he says: 'I have long thought that these consequences of theory, contrary to Newton's principle, would end some day by being abandoned, and yet the recent experiments on the movement of the electrons issuing from radium seem rather to confirm them' (1904, p. 102). A few paragraphs later, he specifies that these are 'the experiments of Kaufmann' (p. 103).[1] Walter Kaufmann conducted an experimental investigation of the mass of high-speed electrons (or cathode rays, as they were still widely called) emitted by radium salts. His results were published in the years 1902–3. An attempt could have been made to explain the variation of mass with velocity, which he found, as an electrodynamic phenomenon and so as not applying to Newtonian or mechanical mass. However, Poincaré concluded that the same laws of variation of mass with velocity must apply to mechanical as well as electrodynamic mass. This implied the falsity of the law of conservation of mass (or Lavoisier's principle), which in turn implied the falsity of Newton's laws. As Poincaré himself puts it:

> So the mechanical masses must vary in accordance with the same laws as the electrodynamic masses; they can not, therefore, be constant.
>   Need I point out that the fall of Lavoisier's principle involves that of Newton's? The latter signifies that the centre of gravity of an isolated system moves in a straight line; but if there is no longer a

constant mass, there is no longer a centre of gravity, we no longer know even what this is. This is why I said above that the experiments on the cathode rays appeared to justify the doubts of Lorentz concerning Newton's principle.

From all these results, if they were confirmed, would arise an entirely new mechanics, which would be, above all, characterized by this fact, that no velocity could surpass that of light. (p. 104)

Poincaré's next piece of research was to carry out a mathematical development of this new mechanics. His paper on the subject was submitted in 1905, and published in 1906.

In 1902 Poincaré had argued that the principles of Newtonian mechanics were definitions, or conventions, which would never be invalidated by experiment. Yet by 1904 he had decided in the light of Kaufmann's experiments that Newtonian mechanics needed to be modified, and by 1905 he had developed the mathematics of a new mechanics. There could scarcely be a more striking example of a scientist-mathematician carrying out a brilliant piece of research which contradicted his own philosophical principles. In effect, Poincaré was conservative in his philosophy of science, but revolutionary in his scientific practice.

The exact converse holds in the case of Poincaré's contemporary, Duhem. Duhem, as we shall see in the next chapter, propounded a philosophy of science in which no theoretical assumption is immune from the possibility of revision and modification. Indeed, Duhem explicitly states that Newton's laws of mechanics may be changed in the future. Yet Duhem, far from welcoming Einstein's theory of relativity, denounced it as an aberration of the German mind. So Duhem was progressive in his philosophy of science, but reactionary in his scientific practice. Later on, I will further examine these curious contradictions between philosophical theory and scientific practice that are to be found in the writings of Poincaré and Duhem.

4.5 Poincaré on the Limits of Conventionalism

Since Poincaré extended his conventionalism from geometry to mechanics, we might have expected him to extend it further to the remaining branches of science. Indeed, a contemporary French philosopher, Le Roy, did produce a global conventionalism of this sort, but Poincaré did not follow in his footsteps. For Poincaré, in fact, most of the remaining laws of science are not conventions or disguised definitions, but genuinely empirical laws founded on induction from observation and experiment. In the following passage, Poincaré states this limit to his conventionalism:

> Principles are conventions and definitions in disguise. They are, however, deduced from experimental laws, and these laws have, so to speak, been erected into principles to which our mind attributes an absolute value. Some philosophers have generalised far too much.[2] They have thought that the principles were the whole of science, and therefore that the whole of science was conventional. This paradoxical doctrine, which is called Nominalism, cannot stand examination. How can a law become a principle? It expressed a relation between two real terms, A and B; but it was not rigorously true, it was only approximate. We introduce arbitrarily an inter-mediate term, C, more or less imaginary, and C is *by definition* that which has with A *exactly* the relation expressed by the law. So our law is decomposed into an absolute and rigorous principle which expresses the relation of A to C, and an approximate experimental and revisable law which expresses the relation of C to B. But it is clear that however far this decomposition may be carried, laws will always remain. (1902, pp. 138–9)

Regarding experimental laws, as opposed to principles, Poincaré gives a Bayesian inductivist account. Observation and experiment can never make a prediction or a law certain, but experience can, nonetheless, render predictions or laws probable, so that 'Every time that he [a physicist] reasons by induction, he more or less consciously requires the calculus of probabilities' (pp. 183–4). Poincaré accordingly devotes a chapter of *Science and Hypothesis* (the ninth) to a consideration of the calculus of probabilities.

He also places a limit to conventionalism on what could be considered 'the other side'. Although he regards geometry as conventional, he does not extend this account to the whole of mathematics. On the contrary, for arithmetic and hence for analysis, he advocates a modification of Kant's theory of synthetic *a priori* intuition.

The English translation of *Science and Hypothesis* appeared in 1905, and Russell reviewed it in *Mind* in the same year. Poincaré replied to Russell's review the next year, and this was indeed the opening of the controversy between Russell and Poincaré to which we have already referred in chapter 3.[3]

By 1905 Russell was a convinced logicist in the philosophy of mathematics, and, therefore, as we might expect, he criticises Poincaré's Kantian account of arithmetic and analysis. Moreover, in accordance with the views on geometry which he had expressed in his 1897 book, Russell rejects Poincaré's claim that geometry is wholly conventional, and argues instead that experience is needed to decide between the various geometries which are *a priori* possible. As Russell puts it:

> All this shows that matter is arranged by perception in a spatial order which is certainly different from some of the possible orders; and it is only for reasons whose origin is in perception that we select at all from among the orders that are *a priori* possible. And this suffices to prove that geometry is not *wholly* conventional, as M. Poincaré contends. (1905, p. 415)

In addition, Russell comments: 'There is also an interesting but unsatisfactory discussion of probability, whose importance, in inductive proofs, M. Poincaré very justly emphasises' (p. 416). To this Poincaré replied, not unreasonably: 'Mr Russell has not the air of being very satisfied with what I say about probability. I am not very satisfied with it either and I would be happy if Mr Russell had something more satisfying to propose' (1906a, p. 143).

The importance of this exchange is that it may have helped to stimulate the interest of Russell and his followers in probability and induction which we described in chapter 1. If this is so, it is not the only instance of the controversy between Russell and Poincaré proving intellectually fruitful as well as entertaining. It will be remembered that Russell discovered a paradox in logic which vitiated Frege's original version of logicism. Indeed, several other paradoxes came to light at about the same time, and, in order to produce a new and satisfactory version of logicism, Russell had to develop some method for solving these paradoxes. In his 1906 polemic against Russell and logicism (reprinted in a shortened version in his 1908 *Science and Method*), Poincaré argues that the paradoxes only affect Cantorian set theory and mathematical logic – two theoretical constructions which he is prepared to abandon *in toto* (cf. his 1908, p. 189).

Given such an attitude, Poincaré should perhaps have merely *ignored* the paradoxes. But he could not resist the temptation of trying to solve them. Remarking with characteristic *brio*, '*Logistic is no longer barren, it engenders antinomies,*' Poincaré proceeded to sketch his own solution to these antinomies (p. 194). This solution involved the so-called vicious circle principle, and Russell incorporated this principle into his own solution, presented in his 1908 paper 'Mathematical Logic as based on the Theory of Types'.

In conclusion, I would like to stress again that the account just given of Poincaré's conventionalism is based largely on chapters 3, 5, and 6 of his *Science and Hypothesis*, which was published in 1902. A fuller account of Poincaré's philosophical views would have to take account of both earlier and later texts. It is clear that the advent of special relativity in the years 1904–6 caused Poincaré to change his opinions in some respects, but different views are possible as to the magnitude of the change. Giedymin (1982 and 1991) has argued in favour of a considerable degree of continuity in Poincaré's thinking. Giedymin believes that we can construct from Poincaré's writings as a whole a sophisticated conventionalist position which is not undermined by special and general relativity. He expounds and develops this position in the works just cited, which are strongly to be recommended to the reader who wishes to explore Poincaré's philosophy further. I, however, will turn in the next chapter to an examination of some further important philosophical ideas of Poincaré's contemporary, Duhem.

5

The Duhem Thesis and the Quine Thesis

In current writings on the philosophy of science, reference is often made to what is called 'the Duhem–Quine thesis'. Really, however, this is something of a misnomer; for, as we shall see, the Duhem thesis differs in many important respects from the Quine thesis. In this chapter I will expound the two theses in turn and explain how they differ. I will conclude the chapter by suggesting that the phrase 'the Duhem–Quine thesis' could be used to refer to a thesis which combines elements from both the Duhem thesis and the Quine thesis. Some use will be made of this suggestion in the final chapter of the book (chapter 10), in which Popper's falsificationism will be examined in the light of the Duhem–Quine thesis.

5.1 Preliminary Exposition of the Thesis. The Impossibility of a Crucial Experiment

Of Duhem's many significant contributions to the philosophy of science, perhaps the most important was his formulation of what I will call the *Duhem thesis*. With his usual clarity and incisiveness, Duhem states this thesis as a section heading thus:

> *An Experiment in Physics Can Never Condemn an Isolated Hypothesis but Only a Whole Theoretical Group* (1904–5, p. 183).

Later in this section he expounds the thesis as follows:

> In sum, the physicist can never subject an isolated hypothesis to experimental test, but only a whole group of hypotheses; when the

experiment is in disagreement with his predictions, what he learns is that at least one of the hypotheses constituting this group is unacceptable and ought to be modified; but the experiment does not designate which one should be changed. (p. 187)

In order to discuss the Duhem thesis, it will be useful to introduce the notion of an *observation statement*. Observation statements will be considered in more detail in chapters 6 and 7 below. For the moment, however, let us take an observation statement to be a statement which can provisionally be agreed to be either true or false on the basis of observation and experiment.

According to the Duhem thesis, an isolated hypothesis in physics (h, say) can never be falsified by an observation statement, O. As a generalisation covering all the hypotheses of physics, this is somewhat doubtful. Physics does appear to contain some falsifiable hypotheses. Consider, for example, Kepler's first law that planets move in ellipses with the Sun at one focus. Suppose that we observe a large number of positions of a given planet and that these do not lie on an ellipse of the requisite kind. We have then surely falsified Kepler's first law. The schema of falsification can be written, where 'not-h' is short for 'It is not the case that h':

If h, then O, but not-O, therefore not-h. (1)

This uses a logical law called *modus tollens*.

However, the Duhem thesis does apply to some hypotheses, and this creates a difficulty for Popper's falsificationism, which we will examine in chapter 10. Consider, for example, Newton's first law of motion (T_1, say). The arguments of Poincaré given in the last chapter and some further arguments to be given in the next chapter indicate that T_1 is not falsifiable. We cannot find an O such that schema (1) above holds when we substitute T_1 for h.

Newton's full theory (T, say) consisted of three laws of motion (T_1, T_2 and T_3) and the law of gravity, T_4. So T was a conjunction of these four laws ($T = T_1 \& T_2 \& T_3 \& T_4$). Even from T by itself, however, we cannot derive any observable consequences regarding the solar system. To do so, we need to add to T a number of auxiliary hypotheses: for example, that no other forces but gravitational ones act on the planets, that the interplanetary attractions are small compared with those between the Sun and the planets, that the mass of the Sun is very much greater than that of the planets, and so on. Let us call the conjunction of such auxiliary hypotheses which are appropriate in a given case A. We now have the schema:

If T_1 & T_2 & T_3 & T_4 & A, then O, but not-O,
therefore not-(T_1 & T_2 & T_3 & T_4 & A).   (2)

Moreover, from not-(T_1 & T_2 & T_3 & T_4 & A) it follows that at least one of the set (T_1, T_2, T_3, T_4, or A) is false, but we cannot say which one.

As the history of science shows, it is often a very real problem in scientific research to decide which one of a group of hypotheses should be changed. Consider, for example, Adams and Leverrier's discovery of Neptune in 1846. From Newton's theory T together with auxiliary hypotheses, astronomers were able to calculate the theoretical orbit of Uranus (the most distant planet then known). This theoretical orbit did not agree with the observed orbit. This meant that either T or one of the auxiliary hypotheses was false. Adams and Leverrier conjectured that the auxiliary hypothesis concerning the number of planets was in error. They postulated a new planet Neptune beyond Uranus, and calculated the mass and position it would have to have to cause the observed perturbations in Uranus's orbit. Neptune was duly observed on 23 September 1846 only 52' away from the predicted position.[1]

This part of the story is quite well known, but there were some subsequent events which are also relevant to the Duhem thesis. Another difficulty which occupied astronomers at the time concerned the anomalous motion of the perihelion of Mercury, which was found to advance slightly faster than it should do according to standard theory. Leverrier tried the same approach that had proved successful in the case of the Uranus anomaly. He postulated a planet Vulcan nearer to the Sun than Mercury, with a mass, orbit, and so forth which would explain the advance in Mercury's perihelion. However, no such planet could be found.

The discrepancy here is very small. Newcomb in 1898 gave its value as 41.24″ ± 2.09″ per century; that is, less than an eightieth part of a degree per century. However, this tiny anomaly was explained with great success by the general theory of relativity (T′), which Einstein proposed in 1915 as a replacement for Newton's theory, T. The value of the anomalous advance of the perihelion of Mercury which followed from the general theory of relativity was 42.89″ per century – a figure well within the bounds set by Newcomb. We see that, although the Uranus anomaly and the Mercury anomaly were *prima facie* very similar, success was obtained in one case by altering an auxiliary hypothesis, in the other by altering the main theory.

In the next section, Duhem goes on to draw an important consequence from his thesis. This section is in fact headed '*A "Crucial Experiment" Is Impossible in Physics*' (1904–5, p. 188). Duhem uses the term *crucial experiment* in something like the sense given by Bacon in the *Novum Organum* to his '*fact of the cross*'. He formulates this notion of crucial experiment as follows: 'Enumerate all the hypotheses that can be made to account for this group of phenomena; then, by experimental contradiction eliminate all except one; the latter will no longer be a hypothesis, but will become a certainty' (ibid.). However, there is an obvious objection to crucial experiments in this strong sense: namely, that we can never be sure that we have listed all the hypotheses capable of explaining a group of phenomena. Duhem makes this point as follows:

> Experimental contradiction does not have the power to transform a physical hypothesis into an indisputable truth; in order to confer this power on it, it would be necessary to enumerate completely the various hypotheses which may cover a determinate group of phenomena; but the physicist is never sure he has exhausted all the imaginable assumptions. (p. 190)

In view of this difficulty, it seems desirable to adopt a rather weaker sense of crucial experiment, which may be defined as follows. Suppose we have two competing theories T_1 and T_2. An experiment (E, say) is crucial between T_1 and T_2, if T_1 predicts that E will give the result O and T_2 predicts that E will give the result not-O. If we perform E, and O occurs, then T_2 is eliminated. If we perform E, and not-O occurs, then T_1 is eliminated. In any event, one of the two theories will be eliminated by E, which is thus crucial for deciding between them. It does not of course follow that the successful theory is necessarily true, because there may be some, as yet unthought of, theory, T_3, which differs from T_1 and T_2 but explains the whole matter much more satisfactorily.

Duhem's point is that if T_1 and T_2 are such that his thesis applies to them, then we cannot derive O from T_1 but only from T_1 and A, where A is a conjunction of auxiliary assumptions. So, if not-O is the result of the experiment, this does not demonstrate beyond doubt that T_1 should be eliminated in favour of T_2. It could be that one of the auxiliary hypotheses in A is at fault.

Duhem illustrates this by what is perhaps the most famous example of an alleged crucial experiment in the history of science: Foucault's experiment, which was designed to decide between the wave theory and the particle theory of light. The wave theory of light predicted that the velocity of light in water should be less than its velocity in air, whereas the particle theory predicted that the velocity of light in water should be greater than its velocity in air. Foucault devised a method for measuring the velocity of light in water, and found that it was actually less than the velocity of light in air. Here, then, we seem to have a crucial experiment which decides definitely in favour of the wave theory of light Indeed, some of Foucault's contemporaries, notably Arago, did maintain that Foucault's experiment was a crucial experiment in just this sense.

Duhem pointed out, however, that to derive from the particle theory that the velocity of light in water is greater than its velocity in air, we need, not just the assumption that light consists of particles (the fundamental hypothesis of the particle theory), but many auxiliary assumptions as well. The particle theory could always be saved by altering some of these auxiliary assumptions. As Duhem puts it: 'For it is not between two hypotheses, the emission and wave hypotheses, that Foucault's experiment judges trenchantly; it decides rather between two sets of theories each of which has to be taken as a whole, i.e. between two entire systems, Newton's optics and Huygens' optics' (p. 189). So, according to Duhem, Foucault's experiment is not a crucial experiment in a strictly logical sense. Yet, as we shall see in the next section, there is another, weaker sense in which the experiment is crucial, even for Duhem.

5.2   Duhem's Criticisms of Conventionalism. His Theory of Good Sense (*le bon sens*)

Duhem is sometimes classified as a conventionalist as regards his philosophy of science, but he is certainly not a conventionalist in the sense of Le Roy and Poincaré. Indeed, he devotes two sections of his *Aim and Structure of Physical Theory* to criticising these thinkers very clearly and explicitly. He formulates their conventionalist position as follows: 'Certain fundamental hypotheses of physical theory cannot be contradicted by any experiment, because they constitute in reality *definitions*, and because certain expressions in the physicist's usage take their meaning only through them' (p. 209).

Duhem objects strongly to Poincaré's claim that the principles of Newtonian mechanics will never be given up, because they are the simplest conventions available and cannot be contradicted by experiment. According to Duhem, the study of the history of science makes any such claim highly dubious:

> The history of science should show that it would be very imprudent for us to say concerning a hypothesis commonly accepted today: 'We are certain that we shall never be led to abandon it because of a new experiment, no matter how precise it is.' Yet M. Poincaré does not hesitate to make this assertion concerning the principles of mechanics. (p. 212; I have here slightly altered the standard English translation in the interests of clarity.)

Poincaré's mistake, according to Duhem, was to take each principle of mechanics singly and in isolation. It is indeed true that when a principle of mechanics – for example, Newton's first law of motion – is taken in this fashion, it cannot be either confirmed or refuted by experience. However, by adding other hypotheses to any such principle, we get a group of hypotheses which can be compared with experience. Moreover, if the group in question is contradicted by the results of experiment and observation, it is possible to change any of the hypotheses of the group. We cannot say with Poincaré that certain fundamental hypotheses, because they are appropriately simple conventions, are above question and can never be altered. This is how Duhem puts the matter:

> It would be absurd to wish to subject certain principles of mechanics to *direct* experimental test; . . .
> Does it follow that these hypotheses placed beyond the reach of direct experimental refutation have nothing more to fear from experiment? That they are guaranteed to remain immutable no matter what discoveries observation has in store for us? To pretend so would be a serious error.
> Taken in isolation these different hypotheses have no experimental meaning; there can be no question of either confirming or contradicting them by experiment. But these hypotheses enter as essential foundations into the construction of certain theories of rational mechanics . . . these theories . . . are schematisms intended essentially to be compared with facts.
> Now this comparison might some day very well show us that one of our representations is ill-adjusted to the realities it should picture, that the corrections which come and complicate our schematism do not produce sufficient concordance between this schematism and the facts, that the theory accepted for a long time without dispute should be rejected, and that an entirely different theory should be constructed on entirely different or new hypotheses. On that day some

one of our hypotheses, which taken in isolation defied direct experimental refutation, will crumble with the system it supported under the weight of the contradictions inflicted by reality on the consequences of this system taken as a whole. (pp. 215–16)

Thus Duhem's position seems to me more accurately described as *modified falsification*, rather than *conventionalism*. Duhem claims that some hypotheses of physics, when taken in isolation, can defy direct experimental refutation. He is thus not a strict falsificationist. On the other hand, he denies that such a hypothesis is immune from revision in the light of experimental evidence. A hypothesis of this kind may be tested indirectly if it forms part of a system of hypotheses which can be compared with experiment and observation. Further, such a hypothesis may on some occasion 'crumble with the system it supported under the weight of contradictions inflicted by reality'. Duhem does not deny that 'among the theoretical elements . . . there is always a certain number which the physicists of a certain epoch agree in accepting without test and which they regard as beyond dispute' (p. 211). However, he is very concerned to warn scientists against adopting too dogmatic an attitude towards any of their assumptions. His point is that, in the face of recalcitrant experience, the best way forward may be to alter one of the most entrenched assumptions. As he says:

Indeed, we must really guard ourselves against believing forever warranted those hypotheses which have become universally adopted conventions, and whose certainty seems to break through experimental contradiction by throwing the latter back on more doubtful assumptions. The history of physics shows us that very often the human mind has been led to overthrow such principles completely, though they have been regarded by common consent for centuries as inviolable axioms, and to rebuild its physical theories on new hypotheses. (p. 212)

Duhem gives as an example the principle that light travels in a straight line. This was accepted as correct for hundreds – indeed, thousands – of years, but was eventually modified to explain certain diffraction effects.

Duhem even cites Newton's law of gravity as a law which is only provisional and may be changed in future. Unfortunately this passage has been accidentally omitted from the English edition of the *Aim and Structure of Physical Theory*. It is here translated from the French edition:

Of all the laws of physics, the one best verified by its innumerable consequences is surely the law of universal gravity; the most precise observations on the movements of the stars have not been able up to now to show it to be faulty. Is it, for all that, a definitive law? It is not, but a provisional law which has to be modified and completed unceasingly to make it accord with experience. (p. 267)

The episode of the anomalous motion of the perihelion of Mercury fits Duhem's analysis perfectly. It would surely have seemed reasonable to explain such a small discrepancy between Newton's theory and observation by altering some auxiliary assumption. In fact, however, the anomaly was only explained satisfactorily when Newton's whole theory of gravity was replaced by Einstein's general theory of relativity. Indeed, from a logical point of view, Duhem's philosophy of science can be seen as offering support to the Einsteinian revolution in physics. It therefore comes as a surprise to discover that Duhem rejected Einstein's theory of relativity in the most violent terms. In his 1915 booklet *La Science allemande* ('German Science'), Duhem argues that Einstein's theory of relativity must be considered as an aberration due to the lack of sound judgement of the German mind and its disrespect for reality. Admittedly, this booklet was written at a time when bitter nationalistic feelings were being generated by the First World War. Indeed, it belongs to a genre known as 'war literature', and is actually a relatively mild example of this unfortunate species of writing. All the same, it is clear that Duhem did reject Einstein's theory of relativity in no uncertain terms.

So, as already observed, we find in both Duhem and Poincaré a contradiction between their philosophical views and their scientific practice. Duhem was led by philosophical considerations to the conclusion that Newtonian mechanics is provisional and may be altered in future; yet he repudiated the new Einsteinian mechanics.[2] Conversely, Poincaré suggested in his philosophical writings of 1902 that the principles of Newtonian mechanics were conventions so simple that they would never be given up; yet, only two years later, in 1904, he decided that Newtonian mechanics needed to be changed, and started work on the development of a new mechanics. Some light is thrown on these strange contradictions by one further element in the Duhem thesis which we have still to discuss. This is Duhem's theory of good sense (*le bon sens*).

Let us take the typical situation envisioned by the Duhem thesis. From a group of hypotheses, $\{h_1 \ldots h_n\}$, say, a scientist has deduced O. Experiment or observation then shows that O is false. It follows that at least one of $\{h_1 \ldots h_n\}$ is false. But which one or ones are false? Which hypothesis or hypotheses should the scientist try to change in order to re-establish the agreement between theory and experience? Duhem states quite categorically that logic by itself cannot help the scientist. As far as pure logic is concerned, the choice between the various hypotheses is entirely open. The scientist in reaching his decision must be guided by what Duhem calls 'good sense' (le bon sens):

> Pure logic is not the only rule for our judgements; certain opinions which do not fall under the hammer of the principle of contradiction are in any case perfectly unreasonable. These motives which do not proceed from logic and yet direct our choices, these 'reasons which reason does not know' and which speak to the ample 'mind of finesse' but not to the 'geometric mind,' constitute what is appropriately called good sense. (1904–5, p. 217)

Duhem imagines two scientists who, when faced with the experimental contradiction of a group of hypotheses, adopt different strategies. Scientist A alters a fundamental theory in the group, whereas scientist B alters some of the auxiliary assumptions. Both strategies are logically possible, and only good sense can enable us to decide between the two scientists. Thus, in the dispute between the particle theory of light and the wave theory of light, Biot, by a continual alteration and addition of auxiliary assumptions, tenaciously and ingeniously defended the particle theory, whereas Fresnel constantly devised new experiments favouring the wave theory. In the end, however, the dispute was resolved.

> After Foucault's experiment had shown that light travelled faster in air than in water, Biot gave up supporting the emission hypothesis; strictly, pure logic would not have compelled him to give it up, for Foucault's experiment was not the crucial experiment that Arago thought he saw in it, but by resisting wave optics for a longer time Biot would have been lacking in good sense. (p. 218)

This passage in effect qualifies some of Duhem's earlier remarks about crucial experiments. Let us take two theories, T_1 and T_2, which are both subject to the Duhem thesis; that is, which cannot be tested in isolation but only by adjoining further assumptions. In a strictly logical sense, there cannot be a crucial experiment which decides between T_1 and T_2. The good sense of the scientific community can, however, lead it to judge that a particular experiment, such as Foucault's experiment, is in practice crucial in deciding the scientific controversy in favour of one of the two contending theories.

In his 1991 book (particularly chapters 4–6), Martin argues that 'lifelong meditation on certain texts of Pascal shaped many of the most important and difficult features of Duhem's thought' (p. 101). In particular, Duhem's theory of good sense (*le bon sens*) was derived in part from Pascal. Indeed, in the passage introducing *le bon sens*, Duhem quotes part of Pascal's famous saying that the heart has its reasons which reason knows nothing of.[3]

Although Duhem was undoubtedly influenced by Pascal, it is possible to suggest factors of a more personal and psychological nature which may have led him to his theory of scientific good sense. As his writings on philosophy of science show, Duhem was a man of outstanding logical ability; yet, as a physicist, he was a failure. In almost every scientific controversy in which he was involved, he chose the wrong side, rejecting those theories such as atomism, Maxwell's electrodynamics, and Einstein's theory of relativity which were to prove successful and lead to scientific progress. Although Duhem stubbornly defended his erroneous scientific opinions, he must have known in his heart of hearts that he was not proving to be a successful scientist. Yet he must also have been aware of his own exceptional logical powers. This situation could only be explained by supposing that something in addition to pure logic was needed in order to become a successful scientist. Here, then, we have a possible psychological origin of Duhem's theory of scientific good sense: namely, that Duhem saw that good sense is necessary for a scientist precisely because he himself was lacking in good sense. Duhem's rejection of a new theory which agreed so well with his own philosophy of science (that is, Einstein's theory of relativity) is just another instance of that lack of good sense which unfortunately characterized Duhem's scientific career.

Poincaré, by contrast, was one of the great physicists of his generation, and was amply endowed with the scientific good sense which Duhem lacked. The contrast between the two men is particularly evident in their respective discussions of electrodynamics. As we have already remarked, Duhem attacked Maxwell's theory harshly, and advocated the ideas of Helmholtz. Poincaré devotes a chapter (the thirteenth) of his 1902 book to electrodynamics. He begins (pp. 225–38) by discussing the theories of Ampère and

Helmholtz and by mentioning the difficulties which he finds in these theories. Then, on p. 239, he introduces Maxwell's theory with the words: 'Such were the difficulties raised by the current theories, when Maxwell with a stroke of the pen caused them to vanish.' Subsequent developments completely endorsed Poincaré's support for Maxwell, while Helmholtz's ideas on electrodynamics, so strenuously advocated by Duhem, are now remembered only by a few erudite historians of science. It was Poincaré's scientific good sense which led him, contrary to the principles of his own conventionalist philosophy of science of 1902, to a modification of Newtonian mechanics.

Duhem's theory of good sense seems to me correct, but, at the same time, more in the nature of a problem, or a starting-point for further analysis, than of a final solution to the difficulty with which it deals. What factors contribute to forming scientific good sense? Why are some highly intelligent individuals like Duhem lacking in good sense? These are important questions, some of which will be raised again later in the book. In the next section, however, I will turn to a consideration of the Quine thesis.

5.3   The Quine Thesis

In his famous 1951 article, 'Two Dogmas of Empiricism', Quine puts forward, with a reference to Duhem, a thesis which is related to Duhem's. Nonetheless, it seems to me that Quine's thesis is sufficiently different from Duhem's to make the conflation of the two intellectually unsatisfactory.[4] I will next briefly describe the Quine thesis,[5] and explain how it differs from the Duhem thesis.

The first obvious difference between Quine and Duhem is that Quine develops his views in the context of a discussion about whether a distinction can be drawn between analytic and synthetic statements, whereas Duhem does not even mention (let alone discuss) the analytic/synthetic problem.

We have already met two ways of defining an analytic statement. The first was due to Kant, who actually introduced the analytic/synthetic distinction. According to Kant, a statement is analytic if its predicate is contained in its subject. This formulation presupposes an Aristotelian analysis of statements into subject and predicate. It is not surprising that Frege, who rejected Aristotelian logic and introduced modern logic, should have proposed a new way of defining an analytic statement. Frege defines an analytic statement as one which is reducible to a truth of logic by means of explicit definitions. These two ways of defining an analytic statement are both illustrated by the standard example of an analytic statement, namely 'All bachelors are unmarried'. But Quine defines analytic statement in yet a third way. He writes critically of 'a belief in some fundamental cleavage between truths which are *analytic*, or grounded in meanings independently of matters of fact, and truths which are *synthetic*, or grounded in fact' (1951, p. 20). In effect, Quine is here taking a sentence to be analytic if it is true in virtue of the meanings of the words it contains. This is the definition of 'analytic' which is adopted by most modern philosophers interested in the question. Once again it is admirably illustrated by the standard example: S = 'All bachelors are unmarried'. Someone who knows the meanings of 'all', 'bachelors', 'are', and 'unmarried' will at once recognise that S is true, without having to make any empirical investigations into matters of fact. Thus S is analytic.

All this seems very convincing; yet Quine denies that the distinction between analytic and synthetic is a valid one. He writes:

> It is obvious that truth in general depends on both language and extralinguistic fact. The statement 'Brutus killed Caesar' would be false if the world had been different in certain ways, but it would also be false if the word 'killed' happened rather to have the sense of 'begat'. Thus one is tempted to suppose in general that the truth of a statement is somehow analyzable into a linguistic component and a factual component. Given this supposition, it next seems reasonable that in some statements the factual component should be null; and these are the analytic statements. But, for all its a priori reasonableness, a boundary between analytic and synthetic statements simply has not been drawn. That there is such a distinction to be drawn at all is an unempirical dogma of empiricists, a metaphysical article of faith. (1951, pp. 36–7)

The empiricists to whom Quine refers are, of course, the empiricists of the Vienna Circle, especially Carnap. As we have seen, their particular brand of empiricism (logical empiricism) did indeed involve drawing a distinction between analytic and synthetic statements. However, support for the distinction is not confined to some members of the empiricist camp. Kantians too support the distinction, which was indeed introduced by Kant himself.

But what has all this to do with the issues involving Duhem and conventionalism, which we have been discussing? We can begin to build a bridge by observing that the meanings given to sounds and inscriptions are determined purely by social convention. Indeed, the social conventions differ from one language to another. So if a sentence is true in virtue of the meanings of the words it contains (that is, is analytic), it is *a fortiori* true by convention. Thus if a law is analytic, it is true by convention. The converse may not hold, since it is conceivable that a law might be rendered true by a set of conventions which include not just linguistic conventions concerning the meanings of words but also perhaps conventions connected with measuring procedures.

Duhem used his thesis against the claim that a particular scientific law was true by convention. It is now obvious that exactly the same argument could be used against the claim that the law is analytic. Indeed, Quine does argue against the analytic/synthetic distinction along just these lines.[6]

But to carry his argument through, Quine makes a claim (the Quine thesis) which is much stronger than the Duhem thesis. The key difference between the two theses is clearly expressed by Vuillemin as follows: 'Duhem's thesis ("D-thesis") has a limited and special scope not covering the field of physiology, for Claude Bernard's experiments are explicitly acknowledged as crucial. Quine's thesis ("Q-thesis") embraces the whole body of our knowledge' (1979, p. 599).

Duhem does indeed place explicit limitations on the scope of his thesis. He writes: '*The Experimental Testing of a Theory Does Not Have the Same Logical Simplicity in Physics as in Physiology*' (1904–5, p. 180). He thinks that his thesis does not apply in physiology or in certain branches of chemistry, and defends it only for the hypotheses of physics. My own view is that Duhem is correct to limit the scope of his thesis, but wrong to identify its scope with that of a particular branch of science – namely, physics. There are in physics falsifiable laws – for example, Snell's law of refraction applied to glass – whereas physiology and chemistry no doubt contain hypotheses subject to the Duhem thesis. When we return to this question in chapter 10, I will argue that the important thing is to distinguish, within any branch of science, two different types of laws, hypotheses, or theories, which I call level 1 and level 2. Level-1 hypotheses are falsifiable by observation statements, whereas level-2 hypotheses cannot be compared with experience in isolation, but only when they are taken in conjunction with other hypotheses. Duhem is thus only partially justified, inasmuch as physics is 'more theoretical' than most other sciences, and so contains a higher proportion of level-2 hypotheses. For the moment, however, it is not of great importance where exact boundaries are drawn. The crucial point is that Duhem wanted to apply his thesis to some statements and not to others, whereas the Quine thesis is supposed to apply to any statement whatever.

This is closely connected with a second difference between the Duhem thesis and the Quine thesis. Duhem maintains that hypotheses in physics cannot be tested in isolation, but only as part of a group. However, his discussion makes clear that he places limits on the size of this 'group'. Quine, however, thinks that the group extends and ramifies until it includes the whole of human knowledge. Quine writes: 'The unit of empirical significance is the whole of science' (1951, p. 42); and again:

> The totality of our so-called knowledge or beliefs, from the most casual matters of geography and history to the profoundest laws of atomic physics or even of pure mathematics and logic, is a man-made fabric which impinges on experience only along the edges. Or, to change the figure, total science is like a field of force whose boundary conditions are experience. A conflict with experience at the periphery occasions readjustments in the interior of the field. . . . But the total field is so underdetermined by its boundary conditions, experience, that there is much latitude of choice as to what statements to re-evaluate in the light of any single contrary experience. No particular experiences are linked with any particular statements in the interior of the field, except indirectly through considerations of equilibrium affecting the field as a whole. (pp. 42–3)

The Quine thesis is stronger than the Duhem thesis, and, in my view, less plausible. Let us take, as a concrete example, one of the cases analysed earlier. Newton's first law cannot, taken in isolation, be compared with experience. Adams and Leverrier, however, used this law as one of a group of hypotheses from which they deduced conclusions about the orbit of Uranus. These conclusions disagreed with observation. Now the group of hypotheses used by Adams and Leverrier was, no doubt, fairly extensive, but it did not include the whole of science. Adams and Leverrier did not, for example, mention the assumption that bees collect nectar from flowers in order to make honey, although such an assumption might well have appeared in a contemporary scientific treatise dealing with a question in biology. We agree, then, with Quine that a single statement may not always be (to use his terminology) a 'unit of empirical significance'. But this does not mean that 'The unit of empirical significance is the whole of science' (1951, p. 42). A group of statements which falls considerably short of the whole of science may sometimes be a perfectly valid unit of empirical significance.

Another difference between Duhem and Quine is that Quine does not have a theory of scientific good sense. Let us take, for example, Quine's statement: 'Any statement can be held true come what may, if we make drastic enough adjustments elsewhere in the system' (p. 43). It is easy to imagine how Duhem would have reacted to such an assertion when applied to a statement falling under his thesis. Duhem would have agreed that, from the point of view of *pure logic*, one can indeed hold a particular statement – for example, Newton's particle theory of light – to be true, come what may. However, someone who did so in certain evidential situations would be *lacking in good sense*, and indeed *perfectly unreasonable*.

Because Quine does not have a theory of good sense, he cannot give the Duhemian analysis which we have just sketched. Indeed, it is significant that his 1951 article, 'Two Dogmas of Empiricism', is reprinted in a collection entitled *From a Logical Point of View*. Where Quine does go beyond logic, it is towards pragmatism, though Quine's pragmatism is usually mentioned only in passing, rather than elaborated, as in the following passage: 'Each man is given a scientific heritage plus a continuing barrage of sensory stimulation; and the considerations which guide him in warping his scientific heritage to fit his continuing sensory promptings are, where rational, pragmatic' (p. 46).

Although the Duhem thesis is quite clearly distinct from the Quine thesis, it might still be possible – indeed, useful – to form a composite thesis containing some, but not all, elements from each of the two theses. The phrase *Duhem–Quine thesis* could then be validly used to denote this composite thesis. In the last section of this chapter, I will elaborate a suggestion along these lines.

5.4   The Duhem–Quine Thesis

Let us say that the *holistic thesis* applies to a particular hypothesis if that hypothesis cannot be refuted by observation and experiment when taken in isolation, but only when it forms part of a theoretical group. The differences between the Duhem and Quine theses concern the range of hypotheses to which the holistic thesis is applied and the extent of the 'theoretical group' for a hypothesis to which the holistic thesis does apply. In discussing these differences, I have so far sided with Duhem against Quine. There is one point, however, on which I would like to defend Quine against Duhem. Quine, as we have seen, extends the holistic thesis to mathematics and logic. Duhem, however, thought that mathematics and logic had a character quite different from that of physics. Crowe (1990) gives an excellent general account and critique of Duhem's views on the history and philosophy of mathematics. I will here confine myself to a brief account of some views concerning geometry and logic which Duhem expounded in his late work *La Science allemande* ('German Science').

Duhem begins his treatment of geometry with the following remarks:

> Among the sciences of reasoning, arithmetic and geometry are the most simple and, consequently, the most completely finished; . . . .
> What is the source of their axioms? They are taken, it is usually said, from common sense knowledge (*connaissance commune*): that is to say that any man sane of mind is sure of their truth before having studied the science of which they will be the foundations. (1915, pp. 4–5)

Duhem agrees with this point of view. In fact, he holds what in 1915 was a very old-fashioned opinion, that the axioms of Euclid are established as true by common-sense knowledge (*connaissance commune*) or common sense (*le sens commun*) or intuitive knowledge (*connaissance intuitive*). A proposition from which Euclid's fifth postulate can be deduced is that, given a geometrical figure (say a triangle), there exists another geometrical figure similar to it but of a different size. Duhem argues that the intuitions of palaeolithic hunters of reindeer were sufficient to establish the truth of this proposition. As he says:

> One can represent a plane figure by drawing, or a solid figure by sculpture, and the image can resemble the model perfectly, even though they have different sizes. This is a truth which was in no way doubted, in palaeolithic times, by the hunters of reindeer on the banks of the Vézère. Now that figures can be similar without being equal, implies, as the geometric spirit demonstrates, the exact truth of Euclid's postulate. (pp. 115–16)

Naturally enough, this attitude to the foundations of geometry leads Duhem to criticize non-Euclidean geometry, and, in particular, Riemannian geometry. This is what he says:

Riemann's doctrine is a *rigorous algebra*, for all the theorems which it formulates are very precisely deduced from its basic postulates; so it satisfies the geometric spirit. It is not a *true geometry*, for, in putting forward its postulates, it is not concerned that their corollaries should agree at every point with the judgements, drawn from experience, which constitute our intuitive knowledge of space; it is therefore repugnant to common sense. (p. 118)

It is perhaps no accident that the non-Euclidean geometer cited by Duhem (namely, Riemann) was a German; for, as already remarked, *La Science allemande*, written in 1915, was an example of the war literature of the time, designed to denigrate the enemy nationality. Duhem attacks German scientists by claiming that, while they possess the geometric spirit (*l'esprit géométrique*), their theories contradict common sense (*le sens commun*) or *l'esprit de finesse*, which is Duhem's new term for something like his old notion of good sense.

Given this general point of view, it is not surprising that we find Duhem condemning the theory of relativity. He speaks of 'the principle of relativity such as has been conceived by an Einstein, a Max Abraham, a Minkowski, a Laue' (p. 135). Forgetting the contributions of his own compatriot Poincaré, he denounces relativity as a typical aberration of the German mind. As he says:

The fact that the principle of relativity confounds all the intuitions of common sense, does not arouse against it the mistrust of the German physicists – quite the contrary! To accept it is, by that very fact, to overturn all the doctrines where space, time, movement were treated, all the theories of mechanics and physics; such a devastation has nothing about it which can displease German thought; on the ground which it will have cleared of the ancient doctrines, the geometric spirit of the Germans will devote itself with a happy heart to re-building a whole new physics of which the principle of relativity will be the foundation. If this new physics, disdainful of common sense, runs counter to all that observation and experience have allowed to be constructed in the domain of celestial and terrestrial mechanics, the purely deductive method will only be more proud of the inflexible rigour with which it will have followed to the end the ruinous consequences of its postulate. (p. 136)

The development and acceptance of non-Euclidean geometry and relativity have rendered Duhem's attempt to found geometry on common sense untenable. It is surely now more reasonable to extend the holistic thesis from physics to geometry and to say that, in the face of recalcitrant observations, we have the option of altering postulates of geometry as well as postulates of physics. This is, after all, precisely what Einstein did when he devised his general theory of relativity.

The picture is the same when we turn from geometry to logic. We have already quoted (chapter 3, note 4) Duhem as saying that 'There is a general method of deduction; Aristotle has formulated its laws for all time (*pour toujours*)' (p. 58). Yet by 1915 the new logic of Frege, Peano, and Russell had clearly superseded Aristotelian logic. Moreover, Brouwer had criticized some of the standard logical laws, and suggested his alternative intuitionistic approach. Quine writes: 'Revision even of the logical law of the excluded middle has been proposed as a means of simplifying quantum mechanics' (1951, p. 43). Admittedly the new 'quantum logic' has not proved very successful in resolving the paradoxes of microphysics; but there is no reason in principle why a change of this kind should not prove efficacious in some scientific context. In artificial intelligence, non-standard logics (for example, non-monotonic logics) are being devised in order to model particular forms of intelligent reasoning, and this programme has met with some success. Thus it seems reasonable to extend the holistic thesis to include logic as well and to allow the possibility of altering logical laws as well as scientific laws to explain recalcitrant observations.

I am now in a position to formulate what I will call the *Duhem–Quine thesis*, which combines what seem to me the best aspects of the Duhem thesis and the Quine thesis. It will be convenient to divide the statement in two parts.

A   The holistic thesis applies to any high-level (level 2) theoretical hypotheses, whether of physics or of other sciences, or even of mathematics and logic. (A incorporates ideas from the Quine thesis.)

B   The group of hypotheses under test in any given situation is in practice limited, and does not extend to the whole of human knowledge. Quine's claim that 'Any statement can be held to be true come what may, if we make drastic enough adjustments elsewhere in the system' (1951, p. 43) is true from a purely logical point of view; but scientific good sense concludes in many situations that it would be perfectly unreasonable to hold on to particular statements. (B obviously follows the Duhem thesis rather than the Quine thesis.)

In what follows I will use the phrase 'the Duhem–Quine thesis' to denote the conjunction of A and B. The thesis seems to me to be both true and important, and in chapter 10 I will examine what consequences it has for Popper's falsificationism. Now, however, we will turn to the third of our four central themes, and examine the often problematic nature of observation.

Part III
The Nature of
Observation

Part III
The Nature of
Observation

6

Protocol Sentences

Let us call a statement which gives the result of an observation or experiment an *observation statement*. Clearly, observation statements play an essential role in science, and we have therefore to investigate their nature and when and how scientists are justified in accepting them.

One view of observation statements is that they are about the sense-impressions, or sense-data, of a particular observer. Thus an observation statement might be something like: 'Brown. Here. Now.' 'I am having a visual sense-datum as of a table.' 'There is a feeling of warmth.' This view can be called *psychologism*. It has a venerable history going back through Mach to the British empiricists, particularly Berkeley. Psychologism about observation statements normally goes with the view that physical objects are constructions out of sense-data, a doctrine known as *phenomenalism*.

While the Vienna Circle, perhaps under the influence of Mach and Russell, were mainly supporters of psychologism and phenomenalism in the 1920s, a shift occurred around 1930 under the influence of Neurath. Neurath's view was that observation statements should not be about sense-data or sense-impressions, but about physical objects. Thus an observation statement, instead of referring to a visual sense-datum or a feeling of warmth, would refer to a brown table or a coal fire. This view was called *physicalism*. The shift in the Vienna Circle from psychologism to physicalism is described in Frank, 1941, pp. 41–7. Neurath, who, as we have seen, was sympathetic to Marxism, may well have been influenced by Lenin, 1908, which is a sustained attack on the psychologism and phenomenalism of Berkeley and Mach. Be that as it may, ever since the shift in the Vienna Circle, the popularity of psychologism and phenomenalism has declined, and they have few advocates today.

6.1   Carnap's Views on Observation Statements
in the Early 1930s

Carnap was persuaded by Neurath to change from psychologism to physicalism, and, like many converts, he became an extreme advocate of the new view, arguing that physicalism applied even to psychology itself. Now it does not seem too unreasonable that the observation statements of sciences like physics and chemistry should be about physical objects. Should not psychology, however, rely on its subjects' reports of their immediate subjective experiences? A psychologist might, for example, study dreams by asking subjects to report their dreams, as well as by noting his own. Thus for psychology, at least, psychologism might seem to apply. However, in his article of 1932/3, Carnap argues that physicalism is correct even for psychology.

Neurath had introduced the term *protocol sentences* for 'observation sentences', and Carnap correspondingly begins by distinguishing the *protocol language* from the *system language*: 'Of first importance for epistemological analyses are the *protocol language*, in which the primitive protocol sentences (in the material mode of speech: the sentences about the immediately given) of a particular person are formulated, and the *system language*, in which the sentences of the system of science are formulated' (1932/3, pp. 165–6). In terms of this distinction, Carnap formulates the thesis of physicalism as follows:

> To every sentence of the system language there corresponds some sentence of the physical language such that the two sentences are inter-translatable. . . . The various protocol languages thus become sub-languages of the physical language. The *physical language is universal and inter-subjective.* This is the thesis of physicalism.
>
> If the physical language, on the grounds of its universality, were adopted as the system language of science, all science would become physics. Metaphysics would be discarded as meaningless. (p. 166)

One point to notice here is the claim that statements about physical objects are inter-subjective. This is really the crux of the matter. The line of thought is this. Let us first consider the protocol sentences of a science like physics or chemistry. Such sciences are carried on by a community of scientific workers. Now a particular observation or experiment may be carried out by scientist A. However, if the result is to be accepted by the community, it is important that A's result should be capable of being checked by another scientist, B. If A's protocol is about physical objects, it can in principle be checked by B, and so is inter-subjective. If, however, A's protocol is about A's private sensations, then A's protocol cannot be checked by B. Carnap goes even further and claims that A's protocol, in these circumstances, becomes meaningless for B. This is his reason for applying physicalism even to psychology. Let us next see how he states the physicalist thesis in the case of psychology.

Carnap considers the example, 'the sentence P_1: "Mr A is now excited."' (p. 170), and comments:

> P_1 has the same content as a sentence P_2 which . . . asserts the existence of that physical structure (microstructure) of Mr A's body (especially of his central nervous system) that is characterized by a high pulse and rate of breathing, which, on the application of certain stimuli, may even be made higher, by vehement and factually unsatisfactory answers to questions, by the occurrence of agitated movements on the application of certain stimuli, etc. (p. 172)

But why do we have to give this interpretation of the sentence P_1, 'Mr A is now excited'? Would it not be more reasonable to say that P_1 refers to an inner psychological state of Mr A? Carnap argues 'No', because in that case P_1 would be testable only by Mr A himself, and not by another individual, B. Carnap assumes that a sentence P_1 is only meaningful for an individual B, if B can test P_1. So on the psychologistic interpretation, P_1 is meaningless for everyone except Mr A. This is absurd, however, since P_1 is clearly meaningful for people other than Mr A, and so Carnap concludes that P_1 must be interpreted in a physicalist rather than a psychologistic way. The same applies even to 'I am now excited' when uttered by Mr A himself. As Carnap puts it:

> Let us say that psychologist A writes sentence p_2: '(I am) now excited' into his protocol. . . . the view which holds that protocol sentences cannot be physically interpreted, that, on the contrary, they refer to something non-physical (something 'psychical', some 'experience-content', some 'datum of consciousness', etc.) leads directly to the consequence that every protocol sentence is meaningful only to its author. If A's protocol sentence p_2 were not subject to a physical interpretation, it could not be tested by B, and would, thus, be meaningless to B. (pp. 192–4)

This paper of Carnap's may well have influenced Wittgenstein's later formulation of 'the private language argument' – that is, the argument that a private language is impossible (cf. Wittgenstein, 1953, sec. 243f.). Indeed, as we shall see in a moment, Neurath explicitly denies the possibility of a private language.

Carnap buttressed his argument about testability and meaning by a consideration, again similar to some passages in the later Wittgenstein, of how a child learns language. He takes the example (1932/3, p. 196) of a mother who, observing her child to be in the physical state of tiredness, puts him to bed, and says: 'Now you are happy to be in bed.' Later, when the child is tired and is put to bed, he says: 'Now I am happy to be in bed', an expression which he has learnt as a way of describing his physical state.

These arguments of Carnap's do not seem to me to be decisive. It could still be claimed that sentences like 'I am happy' or 'Mr A is excited' refer to subjective inner states, while agreeing that we could not refer, in a public language, to such states unless they were correlated with publicly observable physical conditions and behavioural patterns. We will return to this question later on. Let us next consider Neurath's important 1932/3 paper on 'Protocol Sentences'.

6.2   Neurath's Views on Observation Statements in the Early 1930s

Although Neurath's protocol sentences are about physical objects, they do not have a simple form such as: 'There is a table in the room.' The best way to explain their curious structure is to give Neurath's own example of one:

> A complete protocol sentence might, for instance, read: 'Otto's protocol at 3:17 o'clock: [at 3:16 o'clock Otto said to himself: (at 3:15 o'clock there was a table in the room perceived by Otto)].' (1932/3, p. 202)

Neurath goes on to stress that protocol sentences must contain the name of a person (the observer):

> For a protocol sentence to be complete it is essential that the name of some person occur in it. 'Now joy,' or 'Now red circle,' or 'A red

die is lying on the table' are not complete protocol sentences. They are not even candidates for a position within the innermost set of brackets. For this they would, on our analysis, at least have to read 'Otto now joy,' or 'Otto now sees a red circle,' or 'Otto, now sees a red die lying on the table'. (p. 202)

Of course 'Otto now joy' would not be a suitable candidate on other grounds, since Neurath rejects psychologism.

We come next to a most important point. Protocol sentences, according to Neurath, are corrigible, and may be discarded. Here Neurath's position is very different from the earlier psychologistic view. According to psychologism, a protocol sentence describes the immediate experience of an observer. It is thus completely verified by the observer's immediate experience. So protocol sentences constitute a firm and incorrigible basis on which science can be built up. This ceases to be true once we switch to physicalism. Otto can protocol that there is a table in the room and yet, because he is lying or suffering from a hallucination, there is no table in the room. The scientific community may initially accept his protocol, and then come to reject it. Neurath puts the point as follows:

> In unified science we try to construct a non-contradictory system of protocol sentences and non-protocol sentences (including laws). When a new sentence is presented to us we compare it with the system at our disposal, and determine whether or not it conflicts with that system. If the sentence does conflict with the system, we may discard it as useless (or false), as, for instance, would be done with 'In Africa lions sing only in major scales.' One may, on the other hand, *accept* the sentence and so change the system that it remains consistent even after the adjunction of the new sentence. The sentence would then be called 'true.'
>
> The fate of being discarded may befall even a protocol sentence. No sentence enjoys the *noli mi tangere* which Carnap ordains for protocol sentences. (p. 203)

Neurath, of course, here refers to the earlier, psychologistic Carnap. To reinforce the point about protocol sentences being corrigible, Neurath considers an example which curiously antici-pates the findings regarding the split brain. He imagines a scholar called 'Kalon' who, at the same time, writes with his left hand that there is nothing in the room except a table, and with his right hand that there is nothing in the room except a bird. Under these circumstances, the scientific community would have to discard either or both of the protocols.

This article of Neurath's seems to have had an enormous influence on the subsequent course of philosophy. As we have already
remarked, Neurath explicitly denies that there can be a private
language, and so very probably influenced the later Wittgenstein.
The passage in question is the following: 'In other words, *every*
language *as such* is inter-subjective. The protocol of one moment
must be subject to incorporation in the protocols of the next, just as
the protocols of A must be subject to incorporation in the protocols
of B. *It is therefore meaningless to talk, as Carnap does, of a private
language*' (p. 205).

Neurath's rejection of psychologism leads him to formulate the
famous comparison between scientists and sailors who have to
rebuild their ship at sea: '*There is no way of taking conclusively established pure protocol sentences as the starting point of the sciences*. No *tabula
rasa* exists. We are like sailors who must rebuild their ship on the
open sea, never able to dismantle it in dry-dock and to reconstruct it
there out of the best materials' (p. 201).

This passage is cited by Quine, and seems to have influenced him
in developing his holistic philosophy.

6.3   Popper's Views on Basic Statements in 1934

Let us turn next to Popper's treatment of the question in his *Logic of
Scientific Discovery* (1934). Popper deals with the question in chapter
5, which is entitled 'The Problem of the Empirical Basis'. But he
talks neither of 'observation statements' nor of 'protocol statements', but of 'basic statements'. Popper's book was published
shortly after the papers of Carnap and Neurath which we have just
discussed; and, as we shall see, Popper refers frequently to these
papers.

Popper begins his discussion, however, by mentioning J. F. Fries
(1828–31) and his book *Neue oder anthropologische Kritik der Vernunft*
('New or Anthropological Critique of Reason'). Popper extracts
from this work what he calls Fries's trilemma, and expounds it as
follows:

The problem of the basis of experience has troubled few thinkers so
deeply as Fries. He taught that, if the statements of science are not to
be accepted *dogmatically*, we must be able to *justify* them. If we
demand justification by reasoned argument, in the logical sense, then

we are committed to the view that *statements can be justified only by statements*. The demand that *all* statements are to be logically justified (described by Fries as a 'predilection for proofs') is therefore bound to lead to an *infinite regress*. Now, if we wish to avoid the danger of dogmatism as well as an infinite regress, then it seems as if we could only have recourse to *psychologism*, *i.e.* the doctrine that statements can be justified not only by statements but also by perceptual experience. Faced with this *trilemma* – dogmatism *vs* infinite regress *vs* psychologism – Fries, and with him almost all epistemologists who wished to account for our empirical knowledge, opted for psychologism. In sense-experience, he taught, we have 'immediate knowledge': by this immediate knowledge, we may justify our 'mediate knowledge' – knowledge expressed in the symbolism of some language. And this mediate knowledge includes, of course, the statements of science. (1934, pp. 93–4)

As we might expect, Popper disagrees with Fries and 'almost all epistemologists who wished to account for our empirical knowledge' by rejecting psychologism. Thus Popper is here in agreement with Neurath and the later Carnap. However, it is worth noting that the version of psychologism criticized by Popper is rather stronger than that attacked by Neurath and the later Carnap. These two oppose the view that observation statements describe the immediate experience of a particular individual and are completely justified by those experiences. Popper, however, attacks the view that 'statements can be justified not only by statements but also by perceptual experiences'. Thus Popper rules out the possibility that an observation statement might be even partially justified or partially verified by perceptual experience. This strong thesis is justified by the doctrine that statements can be compared only with statements, so that sentences can be tested only by comparing them with other *sentences* (and not with states of affairs or with experience). As Popper sums it up: 'Experiences can *motivate a decision*, and hence an acceptance or a rejection of a statement, but a basic statement cannot be *justified* by them – no more than by thumping the table' (p. 105).

We have earlier endorsed the anti-psychologism of Neurath and the later Carnap, and have agreed that the observation statements used in science must be inter-subjective in character. Popper's stronger version of anti-psychologism seems more questionable, however. For example, is it true that statements can be compared only with statements? Why can't I compare a statement with some of my perceptual experiences? Let us take the statement 'There is a copy of "The Logic of Scientific Discovery" in this room.' At this very moment I am engaged in comparing this statement with a group of my visual and tactile experiences (and, indeed, am finding that it accords with these experiences).

Then again, consider the view that a basic statement cannot be justified, even partially, by perceptual experiences. Suppose I make the statement S: 'There is a tortoise in my bedroom', and that S is disbelieved by Mr A. It is not in fact true that I have to produce another statement, S', in order to justify S. Without uttering a word, I could lead Mr A to my bedroom and point to something crawling about on the bed. S would then be justified by Mr A's perceptual experience. Of course, this justification would only be partial. It might be that the object in question is really an electronic toy purchased by me with the aim of deceiving Mr A. However, to say that a justification is only partial is not to say that it does not exist at all. For these reasons, then, I cannot accept Popper's stronger version of anti-psychologism.

Let us turn next to some criticisms which Popper makes of Neurath. Popper writes:

> Neurath's view that protocol sentences are not inviolable represents, in my opinion, a notable advance. But apart from the replacement of perceptions by perception-statements – merely a translation into the formal mode of speech – the doctrine that protocol sentences may be revised is his only advance upon the theory (due to Fries) of the immediacy of perceptual knowledge.[1] It is a step in the right direction; but it leads nowhere if it is not followed up by another step: we need a set of rules to limit the arbitrariness of 'deleting' (or else 'accepting') a protocol sentence. Neurath fails to give any such rules and thus unwittingly throws empiricism overboard. For without such rules, empirical statements are no longer distinguished from any other sort of statements. Every system becomes defensible if one is allowed (as everybody is, in Neurath's view) simply to 'delete' a protocol sentence if it is inconvenient. (p. 97)

Popper seems to me to be right here. Some rules are indeed needed to limit the arbitrariness of 'deleting' (or else 'accepting') a protocol sentence. But how can Popper himself provide any such rules? In terms of Fries's trilemma, Popper has rejected both psychologism and dogmatism, and so seems to be left with an infinite regress. Any statement is corrigible, and we can justify S only by producing another statement, S'. S' in turn is corrigible, and can be justified only by another statement, S", and so on.

Popper has the problem of evading this infinite regress, which he does in an interesting and ingenious fashion.

Popper thinks that it is indeed possible to go on testing any particular basic statement (S, say) indefinitely. However, such a procedure would render science impossible, and so scientists have to make a decision and reach an agreement to accept S when it has passed a certain number of tests. This decision has a conventional element, and so basic statements are partly conventional. As Popper says, 'Basic statements are accepted as the result of a decision or agreement; and to that extent they are conventions' (p. 106). The decision about when to stop is not entirely arbitrary, however, and Popper mentions, in particular, that we stop at statements which are particularly easy to test. As he says:

> Any basic statement can again in its turn be subjected to tests, using as a touchstone any of the basic statements which can be deduced from it with the help of some theory, either the one under test, or another. This procedure has no natural end. Thus if the test is to lead us anywhere, nothing remains but to stop at some point or other and say that we are satisfied, for the time being.
>
> It is fairly easy to see that we arrive in this way at a procedure according to which we stop only at a kind of statement that is especially easy to test. (p. 104)

This criterion of stopping at statements which are especially easy to test leads Popper to criticize the form of basic statements suggested by Neurath. Popper's point is that it is usually easier to test a statement, S, of the form 'There is a tortoise in my bedroom', than a statement, S', of the form 'Mr A is observing that there is a tortoise in my bedroom'. Indeed, to test S', we may first have to test S, and then, if S turns out to be true, to test that Mr A is indeed observing the reptile in question. As Popper says:

> There is a widespread belief that the statement 'I see that this table here is white', possesses some profound advantage over the statement 'This table here is white', from the point of view of epistemology. But from the point of view of evaluating its possible objective tests, the first statement, in speaking about me, does not appear more secure than the second statement, which speaks about the table here. (p. 99)

In order to evaluate this interesting disagreement between Neurath and Popper, let me first introduce some terminological conventions. I will use Neurath's term *protocol statement*, or *protocol*, for a statement in which the observer is explicitly specified. Thus a typical protocol would be: 'Mr A in such and such a laboratory at such and such a time observed that S'. I will confine the term *observation statement* to statements which report the result of an observation or experiment without in any way alluding to the observer: for example, Popper's 'This table here is white', although sometimes, for emphasis, I will speak of *impersonal observation statements* to emphasize that the statement does not refer in any way to the person making the observation. The question before us is whether protocols or impersonal observation statements should be preferred as the basic statements of science. Neurath argues for protocols, Popper for observation statements – though, as we shall see later, Popper does qualify his view at one point. I will argue that both types of statement are necessary for science, but that their role is rather different.

It should first be noted that protocols do in fact occur in papers by leading scientists. To demonstrate this, we can quote once again the opening of Fleming's 1929 paper on penicillin. Fleming begins the paper as follows:

> While working with staphylococcus variants a number of culture-plates were set aside on the laboratory bench and examined from time to time. In the examinations these plates were necessarily exposed to the air and they became contaminated with various micro-organisms. It was noticed that around a large colony of a contaminating mould the staphylococcus colonies became transparent and were obviously undergoing lysis. (1929, p. 226)

The last sentence here is a protocol. Admittedly the impersonal form; 'It was noticed that . . .' is used; but, in the context of the paper, this is clearly equivalent to: 'Alexander Fleming noticed that . . .'.

The fact that protocols do occur in leading scientific papers is not by itself decisive, for it could be argued that they are in principle dispensable. To show that this is not so and that protocols are really needed, it will be helpful to return to Popper's claim that basic statements are to some extent conventions. Popper illustrates this by drawing an interesting analogy to trial by jury. As he says, 'By its decision, the jury accepts, by agreement, a statement about a factual occurrence – a basic statement, as it were' (1934, p. 109). The verdict of the jury is then used as the basis for further legal procedures: for releasing the prisoner, for example, or for sending him to gaol. In the same way the jury of scientists accepts, by agreement, an observation statement which then becomes part of the data used to evaluate scientific theories or hypotheses.

This analogy seems to me a good one, and it illustrates very nicely the element of social agreement which goes into the acceptance of an observation statement. The analogy does not, however, in my opinion, lend support to Popper's extreme anti-psychologism. Let us suppose that a jury of scientists agrees to accept a particular observation statement S; then S will, in general, have been partly justified or verified, by the perceptual experiences of at least some of the jury. In the same way, the verdict of a jury in a legal case will, in general, be at least partly justified by the evidence they have heard. Of course, in neither case is the justification complete. Both observation statements and the verdicts of juries are corrigible, and can be rejected as a result of later investigation.

Theoretically, each member of the scientific jury could perform the experiment or observation, and partially verify on the basis of his or her own perceptual experiences the observation statement in question. In practice, however, for reasons of expense, lack of time, and so on, the observation or experiment will be performed by *only a few* members of the scientific jury, and the others will rely on the reports of these active observers or experimenters. This is where the need for protocols arises. Consider Mr X, a member of the scientific jury who has not made a particular observation himself. Mr X has to decide whether the corresponding observation statement (S, say) should be accepted, and, to do so, he has to rely on a number of protocols of the form: 'Mr A observed that S', 'Ms B observed that S', and so on. If Mr X knows that Mr A, Ms B and so on are honest and reliable observers or, at least, that they work in institutions noted for the maintenance of high scientific standards, he will be inclined to accept S provisionally as correct. To form this judgement, however, he will need to consider not just the observation statement S, but the protocol statements 'Mr A observed that S', 'Ms B observed that S', and so on, for his acceptance of S depends partly on his judgement as to the reliability of Mr A, Ms B, and the rest. Moreover, if at some later stage, reasons arise for doubting the truth of S after all, Mr X may want to consider the protocols again to see whether Mr A, Ms B, and so on might have been mistaken.

In fact, Popper qualifies his criticism of Neurath at one point, and hints at a 'combined theory' of the type we have just sketched. Thus, he writes:

We stop at basic statements which are easily testable. Statements about personal experience – *i.e.* protocol statements – are clearly *not* of this kind; thus they will not be very suitable to serve as statements at which we stop. We do of course make use of records or protocols, such as certificates of tests issued by a department of scientific and industrial research. These, if the need arises, can be re-examined. Thus it may become necessary, for example, to test the reaction-times of the experts who carry out the tests (*i.e.* to determine their personal equations). (pp. 104–5)

It might be objected that the use of photographs, tape recordings, and so forth could eliminate the need for protocols by individual scientists. Earlier we quoted Alexander Fleming's protocol: 'It was noticed (by Alexander Fleming) that . . .'. Yet Fleming published a photograph of the penicillin culture-plate in the same paper (see our Plate 1). It could be argued that a member of the scientific jury, Mr X, when considering whether to accept the corresponding observation statement, could rely on the photograph, and forget about Alexander Fleming's personal protocol. It is true that photographs and similar devices do help here; but it would be wrong to think that they eliminate the need for protocols altogether. In order to interpret the photograph in the paper, Mr X has to accept that the photograph is of a culture-plate which had been prepared in such and such a way. Moreover, to accept these statements, he has to rely on the protocols of Alexander Fleming. Science accepts and uses impersonal, inter-subjective observation statements; but to accept such statements, it has to rely ultimately on reports which are based on the personal sensory experiences of individual scientists.

To sum up, then, I have diverged from Popper in allowing some degree of psychologism and in according an important role to protocols as well as to impersonal observation statements. It should be stressed, however, that this is *not* in order to provide science with an incorrigible empirical basis, as was the original intention of psychologism. Individual scientists can very easily misinterpret their sensory experience, and so produce mistaken protocols. Indeed, an impersonal observation statement accepted by the scientific jury will often be more certain than the protocols on which it is based, and an individual protocol may sometimes be corrected in the light of an accepted observation statement. Suppose, for example, that five scientists, Mr A, Ms B, . . . , Dr E, independently perform the same experiment and reach exactly the same conclusion, which is expressed in the statement S. Their five protocols: 'Mr A observed that S', 'Ms B observed that S', . . . are pooled by the scientific community, which comes to accept the impersonal observation statement S. Because of the convergence of independent experiments, S is much more certain than any of the protocols upon which it is based. Now suppose further that a sixth scientist, Mr F, has also carried out the experiment and obtained a result different from S. In the light of the acceptance of S, he may re-examine his experimental procedure to find out whether he made some mistake. Very likely he will discover an error in what he did, and, on repeating the experiment in corrected form, obtain S like the others.

But then again, things might go the other way. Mr F might discover that the result S was obtained only because the others had neglected to take some important precaution. They might be persuaded by him to repeat their experiments with this modification, and the end result could be that everyone agreed on Mr F's statement, S', rather than the original S. I am thus in agreement with Popper that all observation statements are corrigible, and would like to close this chapter by quoting the memorable analogy with which he illustrates this corrigibility:

> The empirical basis of objective science has thus nothing 'absolute' about it. Science does not rest upon solid bedrock. The bold structure of its theories rises, as it were, above a swamp. It is like a building erected on piles. The piles are driven down from above into the swamp, but not down to any natural or 'given' base; and if we stop driving the piles deeper, it is not because we have reached firm ground. We simply stop when we are satisfied that the piles are firm enough to carry the structure, at least for the time being. (p. 111)

7

Is Observation Theory-Laden?

7.1 Duhem's View that All Observation in Physics is Theory-Laden

Let us turn now to the views of Duhem concerning observation and experiment in science. Just as in the case of inductivism, so in the present case of observation, we have found it convenient to break the chronological order and to deal with the Vienna Circle and Popper before Duhem. It should not be forgotten, however, that Duhem was writing several years before these Austrian thinkers.

Duhem's treatment of observation in science, like his critique of inductivism, is of the very highest importance, and must be considered a major contribution to the philosophy of science. It is contained in his *Aim and Structure of Physical Theory* (1904–5), part II, chapter 4, entitled 'Experiment in Physics'. With his usual clarity and incisiveness, Duhem begins by stating his central point as follows:

> *An Experiment in Physics Is Not Simply the Observation of a Phenomenon; It Is, Besides, the Theoretical Interpretation of This Phenomenon.* (1904–5, p. 144)

This view, which has come to be generally accepted by philosophers of science, is now usually formulated as the claim that all observation in physics is theory-laden.

Duhem argues for his position by giving the example of measuring the electrical resistance of a coil:

> Go into this laboratory; draw near this table crowded with so much apparatus: an electric battery, copper wire wrapped in silk, vessels filled with mercury, coils, a small iron bar carrying a mirror. An

observer plunges the metallic stem of a rod, mounted with rubber, into small holes; the iron oscillates and, by means of the mirror tied to it, sends a beam of light to a celluloid ruler, and the observer follows the movement of the light beam on it. There, no doubt, you have an experiment; by means of the vibration of this spot of light, this physicist minutely observes the oscillations of the piece of iron. Ask him now what he is doing. Is he going to answer: 'I am studying the oscillations of the piece of iron carrying this mirror?' No, he will tell you that he is measuring the electrical resistance of a coil. If you are astonished, and ask him what meaning these words have, and what relation they have to the phenomena he has perceived and which you have at the same time perceived, he will reply that your question would require some very long explanations, and he will recommend that you take a course in electricity. (p. 145)

In these circumstances, the physicist might make an observation statement S, such as: 'The resistance of the coil is 2.5 ohms.' However, as Duhem's analysis shows, S is the result of interpreting the nature and movements of many pieces of apparatus, using a group of complicated physical theories. The physicist has to devise a theoretical model of the experimental apparatus and perform a series of calculations about this model. Only then can he conclude from the movement of the light beam on the celluloid ruler that the resistance of the coil is 2.5 ohms. The simple observation statement S is highly theory-laden.

Duhem goes on to observe, again quite correctly, that 'The Theoretical Interpretation of Phenomena Alone Makes Possible the Use of Instruments' (p. 153). He argues that this is true even of an instrument as simple as a magnifying glass, and is still more true of a microscope. As he says:

The objects seen through the magnifying glass appear circled by colours of the rainbow; is it not the theory of dispersion which teaches us to regard these colours as created by the instrument, and to disregard them when we describe the object observed? And how much more important this remark is when it is no longer a matter of a simple magnifying glass but of a powerful microscope! (p 154)

So far in Part III of this book we have engaged in a lengthy discussion of the observation statements of science; yet this is the first time that we have had occasion to mention the use of instruments. This is somewhat curious, since virtually all observations in modern science are made with the help of instruments.

However, our present-day reliance on instruments was by no means always a feature of science. Science in the ancient world and during the Middle Ages was based entirely on naked-eye (or, rather, naked human sense-organ) observations. The first really significant use of an instrument for scientific observation occurred when Galileo used a telescope to survey the heavens. Galileo began his observations in 1609, and published his first findings in *The Starry Messenger* (1610). His discoveries were truly remarkable, and showed the great advantage of improving naked-eye observation by the use of an instrument. Galileo was able to observe mountains on the moon with the help of his telescope. He could see at least ten times as many stars as had previously been known, and was able to observe that the Milky Way 'is, in fact, nothing but a congeries of innumerable stars grouped together in clusters' (p. 49). Galileo also discovered that the planet Jupiter has moons circling round it.

Galileo's results were accepted by most people at the time, but some of his Aristotelian opponents questioned whether it was right to rely on observations made with the telescope. Feyerabend has argued that such objections were by no means as absurd and irrational as they might at first appear (1975, pp. 99–144). Galileo had no well-established body of optical theory with which to interpret what he saw through the telescope, and he had to make good this deficiency by a large amount of speculation. Thus Galileo's use of the telescope illustrates both the advantages of instruments and the problems of interpreting theoretically the results they yield.

A further consideration of the use of instruments leads to a criticism of the *physicalism* of Neurath and the later Carnap. Physicalism, it will be remembered, is the doctrine that observation statements are about physical objects. If by 'physical object' here is meant a macro-object such as a table or a chair, the doctrine is certainly wrong. Observation statements can also be about micro-particles too small to be observed with the naked eye. Consider the photograph of a cloud chamber shown in Plate 2. This is interpreted as follows: 'A 63-million-volt positron passes through a 6-mm lead plate and emerges as a 23-million-volt positron.' Here, then, we are observing a positron: a micro-particle of anti-matter of the kind first postulated by Dirac in a high-level physical theory.

In the light of all this, let us consider again Carnap's claim that a sentence such as 'Mr A is now excited' has the same content as a sentence which asserts the existence of a physical structure of Mr A's body, characterized by certain behavioural symptoms. Carnap is driven to this analysis because he doubts the legitimacy of attri-

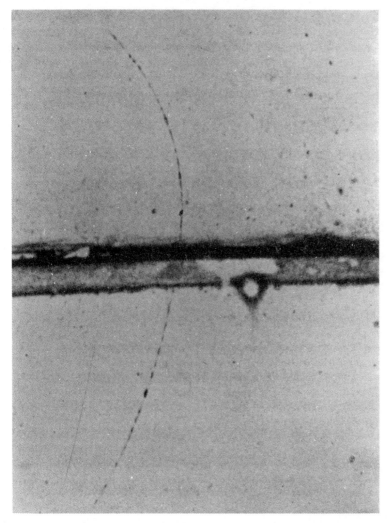

Plate 2 A 63-million-volt positron passing through a 6-mm lead plate and emerging as a 23-million-volt positron. From Max Born, *Atomic Physics* (1935), facing p. 47; reproduced with the permission of the publishers of the current, 8th edition, Dover Publications, Inc.

buting subjective psychological states to Mr A. Yet, if we can interpret the photograph of a cloud chamber as showing a positron passing through a 6-mm lead plate, why should we not interpret Mr A's behaviour as showing that he is in a certain mental state – that of excitement? In both cases we are interpreting observation by means of theory. In the positron case, the theories are those of modern physics. In the case of Mr A, the theories are the traditional theories of common-sense or folk psychology. The attribution of excitement to Mr A is completely inter-subjective. Nearly all observers, armed with the theories of folk psychology, will interpret Mr A's behaviour in the same way, just as nearly all observers who are sufficiently well informed about modern physics will interpret the cloud chamber photograph in the same way. There is actually no need to ask Mr A for introspective reports on his psychological state – though this could be done.

Let us now return to Duhem, who develops his theme by introducing another interesting point. We often speak of eliminating causes of error in an experiment or observation by making appropriate corrections. These corrections, in Duhem's view, consist essentially in improving the theoretical interpretation of the experiment. He illustrates this by the example of the corrections which the French physicist Regnault made to his manometer readings:

> Regnault could represent this real manometer by an ideal one, formed of an incompressible fluid having the same temperature everywhere and subjected at every point of its free surface to an atmospheric pressure independent of the height; between this oversimplified scheme and reality there would be too great a discrepancy and consequently, the experiment would be insufficiently precise. Then he conceives a new ideal manometer, more complicated than the first, but representing better the real and concrete one; he forms this new manometer with a compressible fluid and allows the temperature to vary from one point to another; he also allows the barometric pressure to change when one goes higher up in the atmosphere. All these retouchings of the primitive scheme constitute so many corrections. (p. 157)

Once again, Duhem seems to be right on this point. The corrections which Regnault introduced to obtain better results from his manometer were basically improvements in the theoretical interpretation of his experimental apparatus. Thus, instead of regarding the mercury as an incompressible fluid, he introduced a theoretical model which took account of its compressibility. Instead of regarding the atmospheric pressure on the mercury as independent of its height, he introduced a theoretical model, due to Laplace, which took account of the variation of barometric pressure with height; and so on. Duhem illustrates this process of improving the theoretical interpretation by means of a striking analogy: 'The physicist who complicates the theoretical representation of the observed facts by corrections, in order to permit this representation to come to closer grips with reality, is similar to the artist who, after finishing the line sketch of a drawing, adds shading in order to express better on a plane surface the profile of the model' (p. 158).

As usual, Duhem limits his discussion to physics, but it is clear that his analysis of observation as theory-laden applies just as much to the other sciences as to physics. Let us consider again Alexander Fleming's protocol: 'It was noticed that around a large colony of a contaminating mould the staphylococcus colonies became transparent and were obviously undergoing lysis' (1929, p. 226).

Here various blobs which would convey nothing to the uninstructed lay person (see Plate 1) are interpreted in terms of bacteriological theories involving concepts such as 'staphylococcus colonies' and 'lysis'. Moreover, instruments are now used in all the sciences. We need only think, for example, of the use of the electron microscope in biology. Thus Duhem's points could be illustrated just as well by examples from chemistry or biology as by his own examples from physics.

7.2 A Reinforcement of the Holistic Thesis and Neurath's Principle

The view that all observation in science is theory-laden reinforces the holistic thesis which was formulated in 5.4. We there considered the case of a hypothesis, H_1, which could not be refuted by observation when taken in isolation, but only when taken as part of a conjunction of a group, G, of hypotheses, where $G = \{H_1, H_2, \ldots H_n\}$ say. Now suppose G is refuted by an observation statement, O. This statement, O, is established by the interpretation of sensations in terms of a further group of hypotheses, G', where $G' = \{K_1, K_2, \ldots K_s\}$, say. Thus, to test H_1, we need not only the hypotheses $H_2 \ldots H_n$, but also the hypotheses $K_1 \ldots K_s$. In this sense the holistic thesis is reinforced. To put the same point another way, in the event of the refutation of the group G by O, a scientist has, in addition to the option of changing one or more of the hypotheses in G, the option of querying one of the assumptions in G' in such a way that O is rejected and perhaps replaced by an observation statement, O', which is compatible with G. There are many instances in the history of science where this second strategy was followed with great success.

In 6.2 I quoted Neurath's famous comparison between scientists and sailors who have to rebuild their ship at sea: namely, 'We are like sailors who must rebuild their ship on the open sea, never able to dismantle it in dry-dock and to reconstruct it there out of the best materials' (1932/3, p. 201). This passage comes from Neurath's article on protocol statements, and this is no accident, since the analogy is essentially justified by the point that all observation is theory-laden. To see this, let us first state the content of Neurath's comparison explicitly as a principle which I will call *Neurath's principle*.

Neurath's principle is a conjunction of two parts, (A) and (B), which may be stated as follows:

(A) In order to test any scientific statement, we have to assume for the time being some other scientific statements. (This corresponds in the analogy to the fact that we can only remove one plank of the ship if we leave others in place, since otherwise the ship would sink.)

(B) There is, however, no scientific statement which cannot be subjected to testing, and perhaps abandoned as a result of the tests. (This corresponds in the analogy to the fact that any plank of the ship can be removed and checked to see if it is rotten.)

It is easy to see that both (A) and (B) follow from the view that all observation is theory-laden. As for (A), suppose we are testing a particular scientific statement S, we have to compare S with at least one observation statement, O, and to accept O we need to accept the group of theories (G', say) with which O is 'laden'. Thus, to test S, we need for the time being to accept the scientific statements of G'. As for (B), it follows from the view that all observation is theory-laden that no observation statement O is incorrigible, since the theoretical interpretation involved in O can always be queried. Moreover, such a query can always lead to a further test of O. Thus, any observation statement can be tested and perhaps abandoned. The same is even more obviously true of the non-observation statements of science, and thus (B) follows.

So far, then, I have argued in favour of Duhem's view that all observation in physics is theory-laden, and indeed, have suggested that this view should be extended to other branches of science as well. But we now come to a point where Duhem seems to me to be mistaken. Duhem argues that his view of observation as theory-laden applies to physics, but not to the observation statements of ordinary life. In the next section, however, I will present some findings of empirical psychology which suggest that everyday observation statements are just as theory-laden as those of science. The only difference is that in everyday life the theories applied, usually unconsciously, are common-sense theories shared by all, rather than the high-level theories of a particular branch of science. But let us first examine Duhem's statement of the opposite point of view. He writes:

> An experiment in physics being quite another matter than the mere observation of a fact . . .
>
> When a sincere witness, sound enough in mind not to confuse the play of his imagination with perceptions, and knowing the language he uses well enough to express his thought clearly, says he has observed a fact, the fact is certain: if I declare to you that on such and such a day at such and such an hour I saw a white horse in a certain street, unless you have reasons to consider me a liar or subject to hallucinations, you ought to believe that on that day, at that hour, and in that street, there was a white horse. (1904–5, pp. 158–9)

In legal trials, it is of course important to ascertain whether the observation statements of witnesses are true. The main technique is to cross-examine the witness, and, correspondingly, it is often feared that the witness may be lying. Such fears, according to Duhem, are less justified in the case of physicists, whom he regards as generally trustworthy. Moreover:

> After submitting the physicist's testimony to the rules determining the credibility of a witness's story, we shall have done only a part, the easiest part at that, of the criticism which should determine the value of his experiment.
>
> . . . we must inquire very carefully into the theories which the physicist regards as established and which he used in interpreting the facts he has observed. (p. 159)

There is undoubtedly some truth in what Duhem says here. In legal cases, the main worry is usually whether witnesses are lying. In experimental physics, the main worry is usually whether appropriate theories have been used to interpret the observations. However, the two cases are not quite so sharply distinct as Duhem claims, since the observations of everyday life do seem to be theory-laden. For example, a theoretical interpretation is needed in order to recognize a moving patch of colour as a white horse. This claim is supported by Popper, who takes as an example the statement 'Here is a glass of water':

> Every description uses *universal* names (or symbols, or ideas); every statement has the character of a theory, of a hypothesis. The statement, 'Here is a glass of water' cannot be verified by any observational experience. The reason is that the *universals* which appear in it cannot be correlated with any specific sense-experience. (An 'immediate experience' is *only once* 'immediately given'; it is unique.) By the word 'glass', for example, we denote physical bodies which exhibit a certain *law-like behaviour*, and the same holds for the word 'water'. Universals cannot be reduced to classes of experiences; they cannot be constituted. (1934, pp. 94–5)

Popper is correct here. Suppose, as so often happens in detective stories, that the visitor drains what seems to be an innocuous glass of water and falls dead immediately. What has been interpreted as water was, in reality, a dilute solution of cyanide.

7.3   Some Psychological Findings

The thesis that all observation is theory-laden is strongly supported by some of the investigations of empirical psychologists. Indeed, this point of view has a long history in empirical psychology, and Richard Gregory traces ideas of this sort back to Hermann von Helmholtz's *Treatise on Physiological Optics* (1856–67), in which Helmholtz argues that perceptions are derived by unconscious inferences from sensory signals.[1] Let us start by considering ambiguous figures; in particular, the two most famous such figures: the Necker cube, and the duck-rabbit.

Consider the drawing of a cube in figure 7.1. It is possible to perceive it in two different ways. We can either see the face marked with a's as in front and that marked with b's as behind, or vice versa. With a little practice it is possible to switch at will from seeing the cube in one way to seeing it in the other.

This phenomenon was first described by a Swiss crystallographer, L. A. Necker, in 1832. He noted the effect while trying to draw a crystal which he was viewing through a microscope.

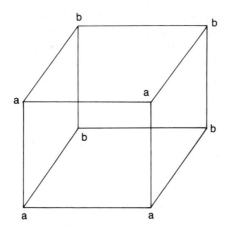

Figure 7.1   The Necker cube

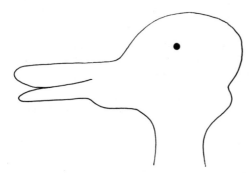

Figure 7.2   The duck-rabbit

The drawing in figure 7.2 can be seen either as a duck or as a rabbit. In the duck interpretation, the two large protuberances are the duck's bill, and the duck is facing left. In the rabbit interpretation, the two large protuberances are the rabbit's ears, and the rabbit is facing right. Once again, it is possible, with a little practice, to switch at will from seeing the drawing as a duck to seeing it as a rabbit, and vice versa.

These ambiguous figures give strong support to the view that all observation is theory-laden. When we look at the duck-rabbit, we do not simply register curved lines and a dot. We see these curved lines either as a duck or as a rabbit. We are providing an active theoretical interpretation of the phenomenon. This interpretation occurs unconsciously, before we become consciously aware of the drawing, and, in this case, two different theoretical interpretations are possible. The lines can be brought under the concept 'duck' or under the concept 'rabbit'. Exactly the same considerations apply to the Necker cube.

Ambiguous figures like the Necker cube and the duck-rabbit have been much discussed (and rightly so) by philosophers as well as psychologists. This is largely due to the influence of Wittgenstein who mentions the Necker cube in his 1921 *Tractatus* (5.5423) and the duck-rabbit in his 1953 *Philosophical Investigations* (II, xi). Following on from Wittgenstein's discussion, Russell Hanson discusses the ambiguous figures in relation to the problem of observation statements in chapter 1 of his 1958 *Patterns of Discovery*.

We have used the ambiguous figures as an argument for the claim that all observation is theory-laden. However, there is an interesting objection to using not just the viewing of ambiguous figures, but the viewing of any figure or picture, as a model for our perception of ordinary objects. The point is put forcibly by the psychologist Richard Gregory in the following passages from his book *The Intelligent Eye*: 'Pictures are in some ways highly artificial inputs for the eye. Although we can learn a lot about perception from pictures, and they are certainly convenient for providing stimulus patterns, they are a very special kind of object which can give quite atypical results' (1970, p. 18).

And again:

Pictures have a double reality. Drawings, paintings and photographs are objects in their own right – patterns on a flat sheet – and at the same time entirely different objects to the eye. We see both a pattern of marks on paper, with shading, brush-strokes or photographic 'grain', and at the same time we see that these compose a face, a house or a ship on a stormy sea. Pictures are unique among objects; for they are seen both as themselves and as some other thing, entirely different from the paper or canvas of the picture. Pictures are paradoxes.

No *object* can be in two places at the same time; no object can lie in two- and three-dimensional space. Yet pictures are both visibly flat and three-dimensional. They are a certain size, yet also the size of a face or a house or a ship. Pictures are impossible. (p. 32)

To avoid this difficulty, it would be as well to give some examples involving the perception just of three-dimensional objects

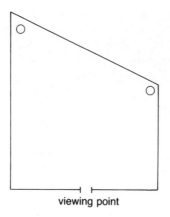
viewing point

Figure 7.3   Plan of the Ames room

which lend support to the thesis that ordinary everyday observation is theory-laden. In fact, Gregory (1970, 1981, and elsewhere) describes quite a number of such examples. We will mention two which are particularly striking and interesting. They are the Ames room and the inside of a mask.

The Ames room was constructed by Adelbert Ames, a man who started as a painter and then turned to creating visual illusions. Figure 7.3 shows the true shape of the room as seen from above. One side of the room is much longer than the other, but the room is designed, according to the principles of perspective, so that from the viewing point, it presents to the eye the same impression as a normal rectangular room.

Now two people of approximately the same size are placed in the room in the corners indicated in the diagram. What the spectator sees from the viewing point is shown in Plate 3. The Ames room is perceived as being of normal shape, while one of the two similarly sized people is seen as very much bigger than the other.

It is easy to understand what is happening here. The brain has the choice of interpreting the visual input according to either of the following two theories:

T_1:   The two people are approximately the same size, but one is much further away because the room is an odd shape.

T_2:   The room is the usual rectangular shape, and the people are different sizes.

Plate 3 The Ames room. From Richard Gregory, *The Intelligent Eye* (1970), p. 26; reproduced with the permission of both the author and the publishers, George Weidenfeld and Nicolson.

Here, T_1 is correct, and T_2 incorrect. Yet anyone, on seeing the Ames room for the first time, will unconsciously opt for T_2. This is perhaps not surprising. Experience has made us all very familiar with the fact that rooms are nearly always rectangular in shape, while people often vary in size. Thus, on the basis of experience, T_2 is better confirmed than T_1, although T_1 is in fact correct. Interestingly enough, if the room is explored with a long stick, it will gradually come to look its true queer shape. Thus further information can alter the way in which the brain unconsciously processes its visual input.

The Ames room shows very clearly that ordinary everyday observation is theory-laden. We are continually interpreting our sensory experience in terms of common-sense theories. Some of these theories may indeed be innate, but others have certainly been learnt by experience. The generalisation that rooms are very likely to be rectangular in shape is something that could only have been learnt by experience. It is very implausible (to say the least) that such a generalisation was genetically programmed into primitive men, who may have lived in the jungle rather than houses.

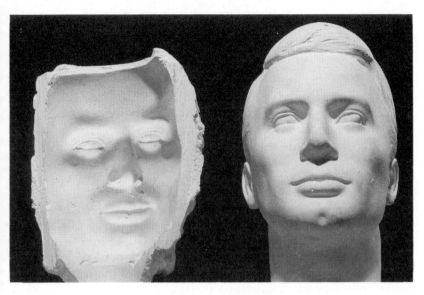

Plate 4 On the left the inside of a mask of the face on the right. Photograph by Richard Gregory, reproduced with his permission.

In Plate 4 the object on the left is the inside of a mask of the face on the right.[2] Although the hollow mask is *viewed from the inside*, it is seen as a face sticking outwards towards the spectator. This is because experience has strongly confirmed for us all that noses stick out of faces rather than into them. We automatically interpret the mask in terms of this well-confirmed hypothesis, unless there is such very strong depth information that the familiar hypothesis is over-ridden, and a less familiar one put in its place. As Gregory says:

> There is nothing impossible about noses sticking inwards. If we look at the back of a mask, close-to in a good light, we see it as a hollow face. There is sufficient depth-information to overcome the improbability of a face being hollow. But when the mask is viewed in a poor light, or an even light casting no shadow or texture, it generally appears as a normal face, though in fact it is inside out. It takes the full power of stereoscopic and texture depth-information from the eyes to convince the visual brain to reject its usual perceptual hypothesis . . . that it is a face with the nose sticking out. (1970, p. 128)

Once again, this example clearly demonstrates that ordinary every-day observation is in fact theory-laden.

7.4 Some General Conclusions

The discussion of an observation statements has been long and complicated, and it is therefore worthwhile trying to summarize a few of the conclusions reached. Let me begin, then, by attempting a definition of an observation statement. The preceding discussion suggests the following: an observation statement is a statement which is the result of some sensory input interpreted, whether consciously or unconsciously, using a set of theories. Instruments may be – indeed, usually are in present-day science – used to produce sensory inputs which would otherwise not have occurred. In that case, theories of the instrument are part of the set of theories used to interpret the sensory input.

The use of ever more complicated instruments can seem to reduce the importance of human sensory input; but this, in fact, always remains ineliminable. Suppose, for example, that a complicated instrument is constructed to observe some process. The instrument is connected to a computer, and the human observer has only to read the computer print-out which states the result of the observa-tion. Even in this extreme case, the human observer has to read the computer print-out, and this involves him or her having some sensory input. Moreover, this input has to be interpreted in accor-dance with a set of theories concerning the process, the instrument, and the computer. So there is a necessary subjective or psycho-logical element in observation; but this is not incompatible, as we shall see in a moment, with observation statements having an inter-subjective character.

This brings us to the topic of *psychologism*. The original version of this theory was that observation statements recount the immediate sensory experiences of a particular individual, and are completely verified by these experiences. This kind of psychologism was quite correctly criticized by Neurath and the later Carnap. There is, however, a weaker version of psychologism, which was attacked by Popper. This weaker version is that an observation statement can be at least partially justified by sensory experience. Popper maintained, on the contrary, that a statement can be justified only by another statement, not by a psychological experience. Here, I sided against

Popper, in favour of weak psychologism. Let S be an observation statement, and E the corresponding sensory experience. There is nothing absurd in claiming that E partially justifies or partially verifies S. Indeed, this seems to me correct. Of course, the justification or verification is only partial, since S involves the theoretical interpretation of E, and this interpretation may well be (indeed, often is) mistaken.

As far as Fries's trilemma is concerned, I have therefore opted for psychologism in its weak form. But the question now arises as to how, in the light of this, it is possible to defend the *inter-subjective* character of observation statements. Neurath and the later Carnap defended inter-subjectivism in the form of physicalism – the view that observation statements must be about physical objects. However, I argued that this was too narrow a framework. Observation statements could be about micro-particles (for example, positrons) or psychological states (for example, anger) as well as about ordinary, macro-, physical objects. For positrons, the sensory input has to be interpreted in terms of the advanced theories of physics. For anger, the sensory input has too be interpreted in terms of the common-sense theories of folk psychology. However, there is no difference in principle between the two cases. Indeed, I argued at length, using examples from empirical psychology, that ordinary everyday observation is just as theory-laden as observation in one of the advanced natural sciences.

Let us go back, therefore, to the following fundamental passage from Neurath: '*Every* language *as such* is inter-subjective . . . the protocols of A must be subject to incorporation in the protocols of B. *It is therefore meaningless to talk, as Carnap does, of a private language*' (1932/3, p. 205). It is by no means necessary to adopt physicalism in order to have this kind of inter-subjectivism. Consider a community of scientists. What is required is, first, that it should be irrelevant which member of the community has the sensory input, I, on which the observation statement, S, is based. So these sensory inputs must have a straightforward character, and not be arcane in any way. One could not use mystical experiences which are accessible only to privileged individuals. One consequence is that a given observation statement is not associated with a single sensory input, I, but rather with a potentially infinite set of such sensory inputs, which could be those of different individuals or of the same individual at different times. A second requirement is that the set of theories used in the interpretation of I should be accepted by all members of the community, and that the application of these theories to the case in hand should have a standard routine character which could be carried out, or at least checked, by any member of the community.

The foregoing analysis of observation statements has many features in common with Kant's analysis of perception (see Kant, 1781/7). Kant believed that conscious perception consists of judgements which are formed by bringing intuitions under concepts. Similarly, I analyse observation statements as formed by giving to sensory input (which corresponds to Kant's intuition) a theoretical interpretation (which corresponds to Kant's 'bringing under concepts'). Of course, there are differences. Kant's intuitions had two pure forms: space and time, whereas I do not postulate that sensory inputs are characterized by any pure forms. Kant's concepts had to include at least one of his twelve pure concepts of the understanding, or categories, whereas I do not claim any such limitation on the character of the theoretical interpretations. Indeed, the theories involved can be of the most varied character, from the theories of common sense to those of an advanced mathematical science such as physics.

The relationship between an observation statement, S, and the corresponding sensory input, I, is a very problematic one, and raises many deep problems concerned with the nature of consciousness and the mind–body problem. We cannot pursue these issues in detail here, but it is only fair to indicate some of the difficulties.

To begin with, it would seem that even the 'rawest' kind of conscious experience has already involved some theoretical interpretation. Thus, pure sensory input has to be considered as something of a construct, or perhaps as occurring unconsciously. If we adopt a physicalist standpoint, we can think of the sensory input as waves or particles – for example, photons – impinging on the human sensory apparatus. This input must already have been considerably processed before it reaches consciousness, and the transformation from electrical excitations in the brain to conscious experience itself raises the whole mind–body problem. This is as far as I will pursue this analysis, and I will now consider another aspect of the problem of observation statements.

Any given observation statement can always be further checked and tested. So Fries is quite right to see here the possibility of an infinite regress. However, scientists are practical people, and the regress has to be cut short by the 'scientific jury' deciding at a certain point to accept a particular observation statement. Popper is right to say that this decision is partly conventional; but I would also stress that it is partly justified (in most cases) by the sensory experiences of at least some of the jury. A decision to accept an observation statement is never final and irrevocable. Observation statements are always corrigible, and Popper is right to see the edifice of science as erected on piles driven into a swamp, rather than on bedrock. None the less the acceptance *pro tem* of some observation statements as data is necessary for the scientific enterprise to go forward.

I will conclude this chapter with an analogy which is constructed by, so to speak, reversing the direction of gravity in Popper's analogy. Popper imagines driving down piles to prevent the weighty edifice of science from sinking into a swamp. Let us instead conceive of scientific theories as hydrogen balloons with a tendency to escape from reality (the earth) into the airy regions of metaphysics. These hydrogen balloons are attached to the earth not by large cables, but by a multitude of fine threads and thin strings, rather like those which held Gulliver captive when he first awoke in Lilliput. Each fine thread is a protocol of the form 'Mr A observed that O'. So each thread represents the sensory experiences of a particular individual interpreted in the light of a set of theories. The thin strings are formed by twisting together a few of the threads. They represent impersonal observation statements, O, which are formed on the basis of individual protocols such as 'Mr A observed that O', 'Ms B observed that O', and so on, but which are more certain than any of the protocols on which they are based, just as the strings are stronger than the threads which compose them. Any thread or string may snap or be cut, but although this may alter the position of the balloon, it will still remain attached to earth because of the multitude of other threads and strings. If we cut *all* the threads and strings, however, our theoretical balloon will float away from the reality of the earth towards the airy regions of metaphysics. Our scientific theory will have become a metaphysical theory. But what exactly is metaphysics? How does it differ from science, and how does it relate to science? These are questions which we shall examine in Part IV of the book.

Part IV
The Demarcation between Science and Metaphysics

8

Is Metaphysics Meaningless? Wittgenstein, the Vienna Circle, and Popper's Critique

8.1 Introduction: The Demarcation Problem and its Importance

We now turn to the fourth theme, which concerns a fundamental problem in the philosophy of science: that of demarcating *scientific* theories from other sorts of theories, particularly *metaphysical* theories. All four themes are of the greatest importance for the analysis of science, but the demarcation problem is, as we shall see, the theme which impinges most on general intellectual concerns outside the realm of science itself. Although the demarcation problem has been much discussed in the twentieth century, it is not new to this century, but was considered by Hume and Kant in the eighteenth century.

As far as the eighteenth century is concerned, the problem was posed by the great successes of the scientific revolution and of Newtonian physics. It seemed to most eighteenth-century thinkers that Newton's theory was a new type of theory, a *scientific* theory, superior in kind to previous theories. At the same time, religion was for the first time in Western Europe coming under heavy attack, partly, no doubt, because of the disillusionment brought about by the wars of religion of the sixteenth and seventeenth centuries. The contrast, then, was between *science*, considered as a sound form of knowledge, and *religious beliefs*, whose claim to be knowledge was more dubious. Hume, an enemy of religion, stated in the famous last paragraph of his *Enquiry concerning the Human Understanding* that volumes of 'divinity or school metaphysics' contain 'nothing but sophistry and illusion' (1748, p. 165).

Kant had a more sympathetic attitude towards religion. He

demarcated religion from science, but thought that, although religion was different from science, it could still be justified up to a point.

These religious problems are by no means dead, as the recent creationist controversy in the USA indicates. Educated common sense has, for a long time, regarded Darwin's theory of evolution as a scientific theory, and the account of the origin of the species in Genesis as a religious myth. The creationists have attacked this viewpoint, claiming that the Genesis account is a theory quite on a par with the Darwinian theory, and that both views should be taught at schools. Can we really say that the Darwinian theory is scientific, and hence superior to the non-scientific account in Genesis? A preliminary study of the demarcation problem is necessary in order to answer this question.

So far, we have considered the issue of science versus religion, but another important question regarding the demarcation problem concerns science versus pseudo-science. There exist certain intellectual activities which are claimed by their proponents to be sciences on a par with the standard natural sciences such as physics, chemistry, and so on, but which are denounced by their detractors as mere pseudo-sciences. The earliest example of such an activity is astrology. Critics of astrology maintained that whereas astronomy was a genuine science, astrology was an illusory pseudo-science.

The issue has become an important one in the twentieth century, because of the emergence of two influential bodies of theory whose status is distinctly doubtful: psychoanalysis and Marxism. The prestige of the natural sciences continued to rise in the nineteenth century with advances in physics, chemistry, and biology. Freud and Marx shared the general enthusiasm for the natural sciences, and both of them aspired to extend science into new areas. Freud saw himself as the founder of a new science of psychology, Marx of a new science of society. Freud compared his discoveries to those of Copernicus and Darwin (1917, p. 351). Engels in his speech at the graveside of Karl Marx in 1883 also made a comparison with Darwin. He said: 'Just as Darwin discovered the law of development of organic nature, so Marx discovered the law of development of human history' (p. 429). Engels, then, considered Marxism to be a scientific analysis of society on which a prediction of the inevitable downfall of capitalism could be securely based. The opponents of Marxism see the matter rather differently. For them, Marxism is a pseudo-science, rather than a genuine science; while the famous prediction of the downfall of capitalism has no scientific basis, and is a piece of mere wish-fulfilment on the part of certain malcontents who are disgruntled with capitalist society. Similarly, opponents of psychoanalysis see Freudian theory as a tissue of myths, rather than a genuine science. Curiously, one of the principal recent opponents of psychoanalysis (Eysenck) has tried to rehabilitate astrology as a science.

Although psychoanalysis and Marxism are the two principal activities whose status, whether science or pseudo-science, is disputed, there are others – for example, intelligence testing. Evans and Waites, in their 1981 book *IQ and Mental Testing*, claim that there is no genuine scientific foundation for intelligence tests, and that the theories about IQ and heredity constitute an 'unnatural science'. Naturally, the advocates of IQ tests would not agree. The issue is clearly one of considerable practical importance, because it concerns the question of whether IQ tests should be used for selection in schools. It is clear that no progress in these disputes is possible without a careful investigation of the demarcation problem, whose importance is thus clearly demonstrated.

The demarcation problem is often formulated as that of distinguishing between science and metaphysics. In this formulation, however, metaphysics must be taken in a broad sense to include both religious doctrines, such as the doctrine of the Trinity, and pseudo-sciences, such as astrology. But metaphysics in this broad sense should also include metaphysics in the narrow sense – that is to say, the general theories produced by philosophers, such as Plato's theory of Ideas, Leibniz's Monadology, and Hegel's account of the Absolute. Philosophers in the past have certainly produced such theories, but should they have done so? Do such theories have any value at all? Or would it be better if intellectuals concentrated on producing theories which were scientific? We will examine these questions as well in the course of our discussion of the demarcation problem.

As far as the twentieth century is concerned, Wittgenstein in his 1921 *Tractatus* and the Vienna Circle concentrated more on distinguishing science from metaphysics in the narrow sense and on the science versus religion aspect of the problem. Popper, on the other hand, devoted more attention to the question of science versus pseudo-science. This is not to say that Popper ignored the other aspects of the problem. Indeed, De Oliveira has argued plausibly in his 1978 article on 'Popper's Two Problems of Demarcation' that in his 1934 *Logic of Scientific Discovery*, Popper, who was closely involved with the Vienna Circle at that time, concentrates more on the distinction between science and metaphysics in the narrow sense, whereas in some of his other writings, particularly his 1963 *Conjectures and Refutations*, he tackles the issue of science versus pseudo-science. These distinctions should be borne in mind; but, as the various aspects of the problem are all interconnected, we will not insist on them too much in what follows.

As we remarked earlier, Russell influenced the Vienna Circle as regards their interest in logic and in the problem of induction. Their interest in the demarcation problem, however, was stimulated by the work of Russell's student Wittgenstein. Indeed, the demarcation problem is one of the central issues in Wittgenstein's first philosophical work, the *Tractatus Logico-Philosophicus* published in 1921. This agrees with Wittgenstein's own opinion, for, writing to Russell about the *Tractatus* on 19 August 1919, he says: 'The main point is the theory of what can be expressed (*gesagt*) by props – i.e. by linguage [*sic*] – (and, which comes to the same, what can be *thought*) and what can not be expressed by props, but only shown (*gezeigt*); which, I believe, is the cardinal problem of philosophy' (Russell, 1968, p. 118). As we shall see, the distinction between what can be expressed and what can only be shown is, within the framework of the *Tractatus*, the distinction between science and metaphysics.

The influence of the *Tractatus* on the Vienna Circle is very well documented. Indeed, Menger, who was invited to join the Vienna Circle in the autumn of 1927 has this to say:

> Only in 1926/27 had the program of the Circle been different, Hahn told me. Various disagreements about Wittgenstein's *Tractatus* had surfaced in the group. Thus, at Carnap's suggestion, Schlick had decided to devote as many consecutive meetings as necessary to read the book aloud. The joint reading, sentence by sentence, filled that entire academic year before I joined. (Menger, 1982, p. 86)

It is thus sensible to start our investigation of the demarcation problem by considering the views of Wittgenstein in the *Tractatus*, and particularly how these views influenced the Vienna Circle.

This is our first encounter with the ideas of Ludwig Wittgenstein, who is certainly one of the most famous philosophers of the twentieth century. In accordance with the usual practice of this book, I will give a few details of his life before discussing his philosophy. It should be added, however, that in the case of Wittgenstein, there are some dangers in this procedure. For Wittgenstein was an unusual man, and led a strange life, which calls out for psychological and sociological analysis. The danger is that of becoming too fascinated by the man, and hence neglecting his philosophy, or perhaps, because of the peculiarities of the man, over-estimating the importance of his philosophy. Yet, so interesting is Wittgenstein's life that I cannot refrain from giving some account of it. I will try, however, to concentrate on those aspects of it which I believe to be relevant to his philosophy. When, later, I come to discuss his philosophy, I will then refer back to the biographical episodes which may have influenced the particular ideas of Wittgenstein under discussion.

8.2 Wittgenstein's Life

Ludwig Wittgenstein was born in Vienna in 1889, and died in Cambridge in 1951.[1] His social background was an unusual one, since his family as one of the richest in Austria, and indeed in Europe. While most of Wittgenstein's relatives were fairly well-to-do, the colossal fortune of Wittgenstein's immediate family was due to the efforts of Wittgenstein's father, Karl. Through a series of remarkable deals, he made himself the leading figure in the steel industry of the Austro-Hungarian Empire, organizing the Austrian Iron Cartel in 1886. Then in 1899, having accumulated a vast fortune, he retired from business at the age of only fifty-two.

Karl Wittgenstein was a patron of the arts as well as a successful entrepreneur. The family residence in Vienna was visited by some of the leading musicians of the day. Gustav Mahler, Bruno Walter, Johannes Brahms, and Clara Schumann frequently came to the Palais Wittgenstein; while Joseph Joachim and the youthful Pablo Casals played in its great salon. Karl Wittgenstein was interested in painting and architecture as well as music. In 1897 a group of progressive artists defected from the conservative Viennese artists' association to form what they called the 'Vienna Secession'. Karl Wittgenstein was their most important benefactor, and gave money for the Secession building which was erected in 1898. This striking building embodied the architectural principles of the group, as well as providing a place where they could hold their exhibitions. The group advocated what was known as the *Jugendstil*, the Viennese version of *art nouveau*. Its most famous painter, Gustav Klimt, did a portrait of Wittgenstein's sister Margarete on the occasion of her marriage in 1905.

So far, the Wittgenstein history is a success story, but the family (in this respect somewhat resembling the Kennedy family) was also to be marked by tragedy. Ludwig was the youngest of eight children, five boys and three girls. Of the five boys, the three eldest committed suicide. The fourth, Paul, also had an unhappy fate. He was training to be a concert pianist, but lost his right arm in the First World War. Ravel wrote the *Concerto for the Left Hand* for him. Ludwig himself suffered from periods of profound depression, and seems to have been near to committing suicide on several occasions. None of the five Wittgenstein brothers married, and at least four of them, including Ludwig, seem to have had homosexual tendencies. Significantly, the three girls were much better adjusted, and two of them married and had children.

Ludwig was at first educated at home by private tutors, but was then sent at the age of fourteen to a grammar school in Linz. Ironically, this school was also attended by Adolf Hitler, who was only a few days older than Wittgenstein. The two were never actually in the same class, since Wittgenstein was a year ahead for his age, whereas Hitler was a year behind. Wittgenstein was not outstanding at school, and in his *Matura* certificate he obtained the highest grade in only one subject: religious knowledge.

After leaving school, Wittgenstein began by studying engineering at the Technische Hochschule in Berlin-Charlottenburg. In 1908 he went to England, and became a research student in the department of engineering at the University of Manchester, where he did research in aeronautics (the design of a propeller). However, his study of the requisite mathematics led him to an interest in the foundations of mathematics, and in 1911 he transferred to Cambridge to study this subject under Bertrand Russell. Wittgenstein's relationship with Russell was decisive for his later work in philosophy.

Russell has the following to say about an early encounter with Wittgenstein:

> At the end of his first term at Trinity, he came to me and said: 'Do you think I am an absolute idiot?' I said: 'Why do you want to know?' He replied: 'Because if I am I shall become an aeronaut, but if I am not I shall become a philosopher.' I said to him: 'My dear fellow, I don't know whether you are an absolute idiot or not, but if you will write me an essay during the vacation upon any philosophical topic that interests you, I will read it and tell you.' He did so, and brought it to me at the beginning of the next term. As soon as I read the first sentence, I became persuaded that he was a man of genius, and assured him that he should on no account become an aeronaut. (1968, p. 99)

The relationship between Russell and Wittgenstein soon became a close one, as is shown by another of Russell's anecdotes:

> He used to come to see me every evening at midnight, and pace up and down my room like a wild beast for three hours in agitated silence. Once I said to him: 'Are you thinking about logic or about your sins?' 'Both', he replied, and continued his pacing. I did not like to suggest that it was time for bed, as it seemed probable both to him and me that on leaving me he would commit suicide. (p. 99)

Russell probably knew that two of Wittgenstein's brothers had by then committed suicide, and this may account for his tolerance of Wittgenstein's strange behaviour. It seems clear that Russell had become an intellectual father figure for Wittgenstein. McGuinness, however, makes the very interesting point that Russell's own emotional state drew him towards 'adopting' Wittgenstein as an intellectual son. McGuinness quotes the following passage from a letter which Russell wrote to Lady Ottoline Morrell on 22 August 1912: 'I had a letter from Wittgenstein, a dear letter which I will show you. I love him as if he were my son.' He then goes on to comment: 'There is no doubt that Russell meant exactly what he said. He was now forty and passionately wanted children but saw no prospects of having any. . . . Wittgenstein to some extent made up for the children he had missed' (McGuinness, 1988, p. 103).

One remarkable feature of Russell's relationship with Wittgenstein is that Wittgenstein came very rapidly to assume the dominating role. In a letter to Lady Ottoline Morrell dated 27 May 1913, Russell has this to say about a recent encounter with Wittgenstein: 'I showed him a crucial part of what I have been writing. He said it was all wrong, not realizing the difficulties – that he had tried my view and knew it wouldn't work. I couldn't understand his objection – in fact he was very inarticulate – but I feel in my bones that he must be right, and that he has seen something I have missed' (quoted from McGuinness, 1988, p. 174). In 1916, in another letter to Lady Ottoline Morrell, Russell recalled this episode as follows:

> Do you remember that . . . I wrote a lot of stuff about Theory of Knowledge, which Wittgenstein criticized with the greatest severity? His criticism, tho' I don't think you realized it at the time, was an event of first-rate importance in my life, and affected everything I have done since. I saw he was right, and I saw that I could not hope ever again to do fundamental work in philosophy. My impulse was shattered, like a wave dashed against a breakwater. I became filled with utter despair. (1968, p. 57)

Russell adds in a footnote: 'I soon got over this mood,' but the fact remains that he did very little fundamental work in philosophy after 1913, devoting himself instead to politics and other interests.

During the First World War, Wittgenstein served with the Austrian army, mainly on the eastern front. It was during this period that he completed his first major philosophical work, the *Tractatus*, which was published in 1921. The circumstances of its composition partly explain why this book, which is largely about logic, language, and meaning, contains quite a number of reflections on death – for example, the following (as is customary we will give quotations from the *Tractatus* by citing the relevant paragraph rather than page number):

> 6.4312   Not only is there no guarantee of the temporal immortality of the human soul, that is to say of its eternal survival after death; but, in any case, this assumption completely fails to accomplish the purpose for which it has always been intended. Or is some riddle solved by my surviving for ever? Is not this eternal life itself as much of a riddle as our present life? The solution of the riddle of life in space and time lies *outside* space and time.
>
> (It is certainly not the solution of any problems of natural science that is required.)

Note how the demarcation problem comes in here. The riddle of our present life is not to be solved by the solution of any problems of natural science, because the riddle belongs to metaphysics.

After the war, Wittgenstein's life entered what was perhaps its most extraordinary phase. His father had died in 1913, but, skilful business man to the last, he had, before his death, transferred virtually all his liquid capital to American stocks and bonds, principally in the United States Steel Corporation (Bartley, 1973, p. 21). Thus the war which ruined so many in Austria greatly increased the Wittgenstein fortune. No sooner had Ludwig Wittgenstein returned to civilian life, however, than he renounced his share of the family money, and arranged for it to be transferred to his siblings. As the author of the *Tractatus* and Bertrand Russell's favourite student, he could, of course, have obtained a university post, and probably a very good one. Yet he renounced this possibility as well, and started a training course for teachers in elementary schools. Then, between 1920 and 1926, he worked as a schoolmaster in three remote Austrian villages (Trattenbach, Puchberg, and Otterthal).

How is this very strange behaviour to be explained? During the war, Wittgenstein had entered a small bookshop in Tarnow in

Galicia. Here he discovered just one book: Tolstoy's *The Gospel in Brief*. He bought it, and read and reread it during the rest of the war. Thus Wittgenstein became converted to what could be called a Tolstoyan Christian populism. I believe that this remained one of his most deeply held beliefs until his dying day.

Tolstoy's Christian populism involves a condemnation of the rich and praise for the virtue and nobility of the poor, particularly of poor peasants. Thus Tolstoy himself says:

> I turned from the life of our circle, acknowledging that ours is not life but a simulation of life – that the conditions of superfluity in which we live deprive us of the possibility of understanding life, and that in order to understand life I must understand not an exceptional life such as ours who are parasites on life, but the life of the simple labouring folk – those who make life – and the meaning which they attribute to it. (1879, p. 67)

We can see that Wittgenstein's actions were very much in accord with this point of view. Indeed, Tolstoy himself had opened a school for the peasants on his estate.

If Wittgenstein had hoped to find noble and virtuous peasants in remote Austrian villages, he was soon disappointed. On 23 October 1921, he wrote to Russell: 'I am still at Trattenbach, surrounded, as ever, by odiousness and baseness. I know that human beings on the average are not worth much anywhere, but here they are much more good-for-nothing and irresponsible than elsewhere' (Russell, 1968, p. 120). It seems that the villagers had an opinion of Wittgenstein which was just about as unflattering as the opinion he held of them.

What had gone wrong? For the answer to this question, we are indebted to the investigations of W. W. Bartley III. In the 1960s Bartley decided to visit Trattenbach and the other villages where Wittgenstein had taught, to see if anyone remembered him. Bartley thought that after more than forty years, Wittgenstein might well have been forgotten, and he took some photographs to jog failing memories. It turned out, however, that nearly everyone of the appropriate age-group remembered Wittgenstein, though no one knew that he had subsequently become a famous philosopher. It seems that Wittgenstein made as big an impact in the Austrian villages as he did at Cambridge or among the Vienna Circle. But the impression seems on the whole to have been unfavourable. Typical descriptions were 'the weird fellow' (Bartley, 1973, p. 86), 'a crazy fellow' (*ein verrückter Kerl*) (p. 84).

At that time a village schoolmaster would normally have lived in a house with at least one servant, and at a standard rather above that of the peasants. The villagers knew that Wittgenstein came from a very rich family, and were therefore most surprised that he chose to live in what Bartley aptly describes as 'ostentatious poverty'(p. 72). Wittgenstein lived in a tiny, primitive room, and regularly had his lunch with one of the poorest families in the area (the Trahts). His evening meal was, if anything, even worse, since it consisted of a mixture of oatmeal and cocoa prepared by himself. It seems that Wittgenstein's eccentricities, which in Cambridge were taken as a sign of genius, were regarded by the Austrian peasants as a sign of mental derangement. The average villager had to work hard to achieve a low standard of living, and no doubt dreamed of improving his condition at least a little. What, then, would he or she think of a rich man who had the means to live well, but lived in poverty for no apparent reason?

The villagers had also found out about Wittgenstein's homosexual tendencies, and this may have been a source of anxiety to the parents of the boys whom he taught. For a long period, Wittgenstein spent every day between 4 p.m. and 7 p.m. giving extra lessons to his favourite pupil, Karl Gruber, whom he even wanted to adopt. Almost certainly this was all quite innocent, but one can understand how such conduct could have alarmed the villagers. When Wittgenstein wanted to take another of his favourite pupils, Oskar Fuchs, to Vienna to see a play at the Burgtheater, Fuchs's mother refused to entrust her son to 'the weird fellow'.

Wittgenstein devoted himself with great intensity to his duties as a teacher. He seems to have been determined that his pupils should reach a high standard and, in particular, learn something about subjects (such as algebra) which were not normally taught in elementary schools. All this proved very successful with a few talented children such as Karl Gruber, but led to conflicts with those of more average abilities. Wittgenstein devoted the first two hours of each day to teaching mathematics, and some of his former pupils recalled remembering these two hours with horror for years afterwards (Monk, 1990, p. 195). There was some resistance on the part of the children to this intensive teaching, and Wittgenstein resorted to corporal punishment. He seems to have been particularly hard on the girls, pulling hair till it came out and boxing ears till the blood ran.

Matters finally came to a head, in Otterthal in April 1926. The incident (cf. Bartley, 1973, pp. 89–90, and Monk, 1990, pp. 232–3) concerned an eleven-year-old boy called Josef Haidbauer. Haidbauer was an unfortunate case. His father was dead, and his mother worked as a live-in maid for a local peasant called Piribauer. Haidbauer was pale and sickly, and later died of leukaemia at the age of fourteen. For some reason, Wittgenstein lost his temper with Haidbauer, and struck him two or three times on the head with such force that the child collapsed. Wittgenstein had previously struck Piribauer's daughter so hard that she bled behind the ears. When Piribauer heard of the Haidbauer incident, he determined on action, and, together with some neighbours, instituted legal proceedings against Wittgenstein. These villagers, who seem to have regarded Wittgenstein as a mentally deranged man with homosexual tendencies, were probably very worried that, if not stopped, he would lose control and hurt one of their children seriously. Wittgenstein was acquitted at the trial, but gave up school teaching thereafter. Ironically, in the autumn of the year in which Wittgenstein's career as a schoolteacher ended so grimly, the Vienna Circle began devoting their seminar to a line-by-line reading of his *Tractatus*.

On his return to Vienna, Wittgenstein worked for a while as a gardener for a monastery, and seriously contemplated becoming a monk. However, his sister found another occupation for him. She had decided to build a large mansion in Vienna, and Wittgenstein and his friend Paul Engelmann became involved in the project. It is not clear how much part Wittgenstein took in the design of the house, but he spent a great deal of time organizing the actual building. According to Bartley, 'he supervised the workmen in the most meticulously careful and exacting way during the construction of the building' (1973, p. 95). By now, however, Wittgenstein was beginning to become involved with philosophy again. Although he never attended a meeting of the Vienna Circle, he started to have philosophical discussions with Schlick and Waismann in 1927. In March 1928 he attended a lecture on the foundations of mathematics by Brouwer, who impressed him deeply. By the summer of 1929, Wittgenstein had decided to return to philosophy on a full-time basis.

As with nearly all aspects of Wittgenstein's life, his return to academia had some distinctly bizarre features. He decided to apply for a Cambridge Ph.D., with the *Tractatus*, by now a world-famous philosophical book, as his thesis. His oral examination was conducted in June 1929 by Russell and G. E. Moore, and Moore is supposed to have presented the following report: 'It is my personal opinion that Mr Wittgenstein's thesis is a work of genius; but, be that as it may, it is certainly well up to the standard required for the Cambridge degree of Doctor of Philosophy' (Bartley, 1973, p. 98). In 1930 Wittgenstein was elected to a fellowship at Trinity College, Cambridge, on the strength of a report written by Russell. Thereafter he remained (with some characteristically Wittgensteinian interruptions, such as a year spent in a hut in Norway) a don at Cambridge, becoming a professor in 1939. He worked as a hospital porter and in a medical laboratory during the Second World War, and then retired from his chair in 1947. He was found to be suffering from cancer in 1949, and died in Cambridge in 1951.

Already by June 1929, Wittgenstein had moved away from the position he advocated in the *Tractatus*, and during his later period at Cambridge, he worked out a completely different philosophy, which was published only after his death. The two main works of the later Wittgenstein are *Philosophical Investigations*, published in 1953, and *Remarks on the Foundations of Mathematics*, published in 1956. These books deal principally with the philosophy of language, the philosophy of psychology, and the philosophy of mathematics. They are thus less relevant to the philosophy of science than the *Tractatus*, which had, because of its influence on the Vienna Circle, a big effect on the development of philosophy of science in the twentieth century. There are, however, some things in the *Philosophical Investigations* which can be related to the demarcation problem, and I will discuss these later in the chapter.

Surveying Wittgenstein's life as a whole, the thing that stands out for me is the contrast between the adulation which he received from elite groups of wealthy intellectuals and the harshness with which he was rejected by the Austrian peasantry. Wittgenstein had no sooner arrived in Cambridge as a young research student than Russell and the other leading philosophers were hailing him as a philosophical genius. When Wittgenstein's sister visited him in Cambridge in 1912, she had tea with Russell, who said to her: 'We expect the next big step in philosophy to be taken by your brother' (McGuinness, 1988, p. 130). The Vienna Circle regarded the *Tractatus* as a turning-point in philosophy. When Wittgenstein returned to Cambridge, he was treated with no less awe and reverence, as Malcolm's memoir (1958) clearly shows.

Curiously, Wittgenstein seems to have treated these elite groups who adulated him with a fair measure of indifference, if not downright contempt. Russell secured Wittgenstein's election to Cambridge's most exclusive secret society, the Apostles – later to become a breeding ground for Russian spies. But Wittgenstein resigned after attending only one meeting. This was at least one more than the number of meetings of the Vienna Circle which he attended. My impression is that Wittgenstein's populism was not an affectation, but was quite genuinely felt; that Wittgenstein really believed the wealthy elite to be degenerate, and that virtue was to be found among the poor rather than the rich. Yet his attempts to mix with the Austrian peasantry ended with their taking legal action to get rid of him. We know nothing of his relations with the workers who built his sister's mansion; but Bartley's description of his supervising them 'in the most meticulously careful and exacting way' (1973, p. 95) suggests that his relations with them may not have been very good either. What moral can be drawn from all this? Perhaps the simple one that it is hard for anyone to escape his or her class background.

So much then for Wittgenstein the man; let us now turn to his philosophy.

8.3 Wittgenstein's *Tractatus*

Let us begin with a number of quotations from the *Tractatus* which show the views on metaphysics held by Wittgenstein at that time (1921).

4.003 Most of the propositions and questions to be found in philosophical works are not false but nonsensical. Consequently we cannot give any answer to questions of this kind, but can only establish that they are nonsensical. Most of the propositions and questions of philosophers arise from our failure to understand the logic of our language.

(They belong to the same class as the question whether the good is more or less identical than the beautiful.)

And it is not surprising that the deepest problems are in fact *not* problems at all.

This point of view leads him to characterize what he regards as the correct method in philosophy as follows:

6.53 The correct method in philosophy would really be the follow-ing: to say nothing except what can be said, i.e. propositions of natural science – i.e. something that has nothing to do with philosophy – and then, whenever someone else wanted to say something metaphysical, to demonstrate to him that he had failed to give a meaning to certain signs in his propositions. Although it would not be satisfying to the other person – he would not have the feeling that we were teaching him philosophy – *this* method would be the only strictly correct one.

For Wittgenstein, then, the correct method in philosophy is to demonstrate that any given metaphysical proposition is meaningless. For this purpose, it would clearly be desirable to have a theory of meaning, and Wittgenstein does indeed provide such a theory in the *Tractatus*. It must be added, however, that later in his life he criticized and gave up his *Tractatus* theory of meaning, and in his second major philosophical work, the *Philosophical Investigations* (1953) advocated a different theory of meaning. My own opinion is that Wittgenstein's self-criticism is correct, and that, although there are some important ideas in his earlier (*Tractatus*) theory of meaning, his later (*Investigations*) theory of meaning is preferable. Now the interesting thing (as I will argue later) is that the doctrine that metaphysics is meaningless does indeed follow from the *Tractatus* theory of meaning, but not from the *Investigations* theory. Let us begin, however, with a brief exposition of the *Tractatus* theory.

This theory is based on the notion of an elementary proposition, which is supposed to assert that a simple fact is the case. As Wittgenstein himself says:

4.21 The simplest kind of proposition, an elementary proposition, asserts the existence of a state of affairs.

All other meaningful propositions are built up out of elementary propositions. We can explain this by an analogy with chemistry. The elementary propositions correspond to atoms, and the non-elementary (or complex) propositions to molecules. Indeed, elementary propositions can also be called *atomic* propositions, and complex propositions *molecular* propositions. The process by which molecular propositions are built up out of atomic propositions is known as *truth-functional composition*. As Wittgenstein says:

5 A proposition is a truth-function of elementary propositions.

By 'proposition' here, Wittgenstein means 'meaningful proposition'. The 'propositions' of metaphysics and even, rather surprisingly, those of mathematics (cf. 6.2) are pseudo-propositions.

In a moment I will explain what is meant by truth-functional composition, but first let us consider a more basic question. Wittgenstein's *Tractatus* theory of meaning is based on the notion of elementary (or atomic) propositions, and we might well expect that he would clarify this notion by giving some examples of elementary propositions. But he never in fact does this; instead, he makes a few general remarks about elementary propositions. We learn, for example, that elementary propositions consist of names, and that names mean objects.

> 4.22 An elementary proposition consists of names. It is a nexus, a concatenation, of names.

> 3.203 A name means an object. The object is its meaning.

Malcolm has an interesting anecdote regarding this:

> I asked Wittgenstein whether, when he wrote the *Tractatus*, he had ever decided upon anything as an *example* of a 'simple object'. His reply was that at that time his thought had been that he was a *logician*; and that it was not his business, as a logician, to try to decide whether this thing or that was a simple thing or a complex thing, that being a purely *empirical* matter! It was clear that he regarded his former opinion as absurd. (1958, p. 86)

Granted this obscurity in the *Tractatus* itself, it was up to the Vienna Circle to clarify the matter, and this they did by identifying elementary propositions with simple observation statements, or *protocol* statements. This identification led in turn to the involved debates about the nature of observation statements which we considered in chapter 6. We shall not go over these subtleties again here, but confine ourselves to such common-sense examples of simple observation statements as 'The apple on my desk is green'. This is because our aim now is not to analyse simple observation statements, but to discuss how complex observation statements can be built up out of simple ones.

Later scholars have argued that the interpretation of 'elementary proposition' adopted by the Vienna Circle is not really compatible with the *Tractatus*, and so probably does not represent what Wittgenstein had in mind. This view seems to have originated in Anscombe's *An Introduction to Wittgenstein's Tractatus* (1959) (see particularly, chapter 1, 'Elementary Propositions', pp. 25–40). Anscombe quotes the following passage from the *Tractatus*:

6.3751   It is clear that the logical product of two elementary propositions can neither be a tautology nor a contradiction. The statement that a point in the visual field has two different colours at the same time is a contradiction.

She then comments:

> If elementary propositions are simple observation statements, it is very difficult to see how what Wittgenstein says here can possibly hold good of them; for, for any proposition which could reasonably be called a 'simple observation statement', one could find another that would be incompatible with it and be precisely analogous to it logically. Therefore, whatever elementary propositions may be, they are not simple observation statements. (1959, p. 27)

This argument is certainly very convincing.

The view that the elementary propositions of the *Tractatus* were not intended to be observation statements is also adopted by Janik and Toulmin in their interesting book *Wittgenstein's Vienna* (1973) (see, for example, pp. 145 and 212–21). They suggest, as an alternative interpretation, that elementary propositions might be statements about the ultimate structure of the world – perhaps statements of the deepest possible kind of theoretical physics. Thus the 'objects' of the *Tractatus* would be more in the nature of elementary particles than of colour patches.

There is in fact considerable evidence in favour of this second interpretation. It fits well with Malcolm's anecdote recounted earlier, and gives a better explanation of the gnomic utterance:

2.0232   In a manner of speaking, objects are colourless.

This seems definitely to rule out the suggestion that 'objects' could be either physical objects or immediate sense-data. On the other hand, it is quite compatible with the view that 'objects' are some kind of physical elementary particles whose combination gives rise to colour. As far as the development of philosophy of science in the twentieth century is concerned, however, what Wittgenstein actually thought when writing the *Tractatus* is of much less importance than how he was interpreted by the Vienna Circle. I will therefore expound what is in effect a hybrid theory of meaning, in which meaningful propositions are regarded as truth-functions of elementary propositions, and elementary propositions are identified with simple observation statements. This hybrid theory, as we shall see, led to the Vienna Circle's famous verifiability criterion.

It is now time to explain what is meant by truth-functional composition, and, in accordance with our programme, we shall do so in terms of observation statements. Let us begin with a simple observation statement O_1 = 'The apple on my desk is green'. Following common sense and ignoring for the moment some of the subtleties considered in Part III, we can say that the truth of this statement can be checked by a straightforward observation. If the statement is true, this can be *verified* by observation. Next consider the negation of O_1 – that is, not-O_1 = 'It is not the case that the apple on my desk is green'. Once again, the truth of this statement can be checked by a straightforward observation, so that, in general, if O is a simple observation statement, then so is its negation, not-O.

Let us now go on to consider complex observation statements. To do so, let us take another simple observation statement, O_2 = 'The apple on the kitchen table is red'. This can also be checked by a straightforward observation. However, we cannot check O_1 and O_2 simultaneously. I can check O_1 by making observations in my study, but then, in order to check O_2, I have to descend the stairs and enter the kitchen, a process which clearly takes time. This suggests that we might characterize a simple observation statement as one whose truth-value can be checked by observations made at a particular time and place. A complex observation statement is then one whose truth-value (whether true or false) can indeed be ascertained by observations, but by observations which have to be made at different places and/or times. An example of a complex observation statement would be: O_1 & not-O_2 = 'The apple on my desk is green, and it is not the case that the apple on the kitchen table is red'. If this is true, its truth can certainly be verified by observation, but two observations at different places and times are needed.

In our earlier discussions, the phrase 'observation statement' was sometimes used to mean simple observation statement, and at other times to cover both simple and complex observations statements. This ambiguity was deliberate, since insisting on the distinction would have complicated the argument unnecessarily. Now that we have introduced the distinction, however, the ambiguity can be cleared up retrospectively. In Part III, 'observation statement' is always used to mean simple observation statement. Thus, at the beginning of chapter 6 I say: 'Let us call a statement which gives the result of an observation or experiment an *observation statement.*' After a good deal of analysis, this preliminary definition is developed into the following (7.4): 'An observation statement is a statement which is the result of some sensory input interpreted, whether consciously or unconsciously, using a set of theories.' Both definitions are clearly appropriate for simple observation statements. In 5.1, however, I say: 'let us take an observation statement, O, to be a statement which can provisionally be agreed to be either true or false on the basis of observation and experiment.' This definition covers both simple and complex observation statements.

I am now in a position to explain what is meant by truth-functional composition – the process by which molecular propositions are built up out of atomic propositions. It should be clear that if O is an observation statement, then so is not-O. The very same observations which determine whether O is true or false determine whether not-O is false or true; since if O is true, not-O is false, and if O is false, not-O is true. Similarly, if O_1, O_2 . . . O_n are a finite set of observation statements, then O_1 & O_2 & . . . & O_n is in general an observation statement. (There is a small exceptional case which we shall mention later.) The molecular proposition O_1 & O_2 & . . . & O_n is true only if all its component atoms O_1, O_2 . . . O_n are true, and is false if any of these atoms are false. But since O_1 . . . O_n are all observation statements, we can determine by observations whether they are true or false, and these same observations determine whether O_1 & . . . & O_n is true or false.

The particles 'not-' and '&' are known as *connectives*; they are, so to speak, the glue which sticks atomic propositions (simple observation statements) together to form molecular propositions (complex observation statements). The procedure can be imagined as follows. Suppose we start with a finite set of simple observation statements. We first form the negations of each of these statements, and add them to the set. We then take any finite subset of statements and form their conjunction; that is, we join them together with the connective '&', and add these statements to the set. We then form the negations of all the statements in the new set, and add these to form a set which is still further extended. We then form the conjunctions of all subsets of this new set, and so on, indefinitely. We can briefly summarize the process by saying that, from a set of atomic propositions, we form an extended set of molecular propositions by means of the connectives 'not-' and '&'.

If the original set consisted of simple observation statements, the extended set will, as we have argued, in general consist of observation statements. There are, however, some exceptional cases which we must now mention. The connectives 'not-' and '&' can sometimes generate a logical contradiction or a logically valid statement instead of an observation statement. For example, if O is any observation statement, then O & not-O is a logical contradiction (always false), while not-(O & not-O) is a logically valid statement (always true). If, however, we take a set of simple observation statements, generate the extended set of statements using the connectives 'not-' and '&', and then eliminate all the logical contradictions and logically valid statements, we are left with a set consisting entirely of observation statements.

As a matter of fact, we can get the same extended set by using connectives other than 'not-' and '&', and Wittgenstein in the *Tractatus* used a single connective known as the *Sheffer stroke symbol*. The general notion involved here is that of a *truth-function*. We say that a proposition P is a truth-function of propositions $P_1, P_2 \ldots P_n$ if the truth-value of P (whether P is true or false) is determined by the truth-values of $P_1, P_2 \ldots P_n$.

If we start with a set of atomic propositions, it can be shown quite easily that the connectives 'not-' and '&' will generate a set consisting precisely of the truth-functions of these atomic propositions. The central thesis of the hybrid theory which we are considering (that is, Wittgenstein's *Tractatus* theory as interpreted by the Vienna Circle) is that all meaningful propositions are truth-functions of simple observation statements. It follows that any meaningful proposition is either a logical contradiction or a logically valid statement or an observation statement. It further follows that if a meaningful proposition is true, this can be verified by observation. The meaningful proposition (P, say) is, by the central premise of the theory, a truth-function of simple observation statements. We can determine by observation the truth-values of these simple observation statements, and in this way we verify that P is true. This leads to the Vienna Circle's famous *verifiability criterion*.

It is easy to see that, on this account of meaning, metaphysics becomes meaningless. But the theory has other consequences as well. For instance, all ethical propositions become meaningless. Wittgenstein draws this conclusion quite explicitly:

6.42   And so it is impossible for there to be propositions of ethics.
       Propositions can express nothing that is higher.
6.421 It is clear that ethics cannot be put into words.
       Ethics is transcendental.
       (Ethics and aesthetics are one and the same.)

Later on, for essentially the reasons given by Popper, I will reject the central premise of the hybrid theory: namely, that all meaningful propositions are truth-functions of simple observation statements. None the less, I believe that there is a valuable idea here which can be retained. I think that it is correct to say that *all observation statements* are truth-functions of simple observation statements, and I will call this the *Tractatus*/Vienna Circle characterization of the class of observation statements. The thesis is that the *Tractatus*/Vienna Circle theory does correctly characterize the class of observation statements, and that the Vienna Circle's mistake was to identify this class with the class of all meaningful statements. They had far too narrow a conception of what was meaningful.

8.4   The Vienna Circle on Metaphysics

The Vienna Circle accepted Wittgenstein's view that metaphysics is entirely meaningless. The classic expression of their point of view is to be found in Carnap's 1932 article 'The Elimination of Metaphysics through Logical Analysis of Language'. Here Carnap writes emphatically that 'In the domain of *metaphysics*, including all philosophy of value and normative theory, logical analysis yields the negative result *that the alleged statements in this domain are entirely meaningless.* Therewith a radical elimination of metaphysics is attained' (1932, pp. 60–1). This is not to say that Carnap's views are identical to Wittgenstein's. In fact, as we shall see later on, Carnap changed and developed the ideas of the *Tractatus*. In particular, Carnap used various kinds of logical apparatus which were more complicated than Wittgenstein's truth-functional composition. I will not go into these technical details here, however, partly because these additional complications do not affect the basic philosophical argument and partly because I think that truth-functional composition does in fact suffice to characterize the class of observation statements.

In his 1932 paper, Carnap states the verifiability criterion as follows: 'The meaning of a statement lies in the method of its verification. A statement asserts only so much as is verifiable with respect to it. Therefore a sentence can be used only to assert an empirical proposition, if indeed it is used to assert anything at all' (p. 76). He selects some passages from Heidegger's *Was Ist Metaphysik?* (1929), and then proceeds to demonstrate to his own satisfaction that they are meaningless. Here are some of the

Heideggerian pronouncements with which Carnap deals: 'Where do we seek the Nothing? How do we find the Nothing? . . . We know the Nothing. . . . *Anxiety reveals the Nothing.* . . . That for which and because of which we were anxious, was 'really' – nothing. Indeed: the Nothing itself – as such – was present. . . . *What about this Nothing? – The Nothing itself nothings'* (cf. Carnap, 1932, p. 69; as Carnap points out, the emphasis is in the original).

The thesis that these statements are meaningless is by no means implausible. With a certain ponderousness which is not lacking in humour, Carnap tries to translate Heidegger's propositions into a precise logical language, and concludes that sentences like '*The Nothing itself nothings'* cannot even be expressed in such a language. Still less can we specify the finite set of simple observation statements which would verify the claim that *the Nothing itself nothings,* if, indeed, it were true. It follows, according to Carnap, that '*The Nothing itself nothings'* is meaningless – a pseudo-proposition.

In the final section of his paper, Carnap considers another defence of metaphysics: namely, that it serves for 'the *expression of the general attitude of a person towards life'* (p. 78). This view is not so implausible. After all, feelings of intense anxiety such as Heidegger expresses were, as it turned out, eminently justified in the Germany of 1929. Yet Carnap is not prepared to admit even this defence. It is quite legitimate for someone to express his or her basic attitude to life, but this should be done through art (lyrical poetry, music, and so forth), not through metaphysics. The problem with the metaphysical mode of expression of a basic attitude is that it deludes the metaphysician into thinking that 'he travels in territory in which truth and falsehood are at stake', whereas, 'In reality . . . he has not asserted anything, but only expressed something, like an artist' (p. 79). Thus, the metaphysician foolishly tries to argue for his position, to refute the views of his opponents, while the lyrical poet – quite correctly, because he is simply expressing an attitude to life – refrains from doing any of these things. Carnap concludes that 'Metaphysicians are musicians without musical ability' (p. 80), and praises Nietzsche for writing *Thus Spake Zarathustra* in the form of poetry rather than traditional metaphysics.

It is not difficult to detect a political motivation in the Vienna Circle's attack on metaphysics. The political views of the Circle were, broadly speaking, liberal and left liberal, and their principal opponents were the Catholic reactionary parties. They surely then had a reason for wanting to demonstrate that Catholic theology was meaningless verbiage. In the secular sphere, they would naturally find themselves combatting the leading philosophers of the authoritarian right, who then, as now, were Heidegger and Nietzsche. Heidegger gave his support to Nazism, and Nietzsche was Hitler's favourite philosopher. It is understandable that a liberal such as Carnap would want to dismiss the writings of these philosophers as either meaningless or mere emotional effusions. Yet the very success and social influence of Catholic theology and of the philosophies of Heidegger and Nietzsche cast doubt on Carnap's viewpoint. It may not be easy or, indeed, possible to express Heidegger's writings in the precise language of formal logic, but we should not therefore conclude that Heidegger's philosophy is altogether meaningless.

Wittgenstein's attitudes were somewhat different from those of Carnap and most other members of the Vienna Circle. Wittgenstein certainly, as we have seen, held the view that metaphysics was meaningless. He was, however, a man of strong religious leanings, even though not a member of any organized religion; and in the *Tractatus* he developed a theory of the mystical. The key point here is that, for Wittgenstein in the *Tractatus*, the limits of what can be meaningfully said do not coincide with the limits of what can be thought. On the contrary, there are things which cannot be said, but which can none the less be shown or thought or which make themselves manifest. Such things constitute the mystical. As Wittgenstein himself puts it in the preface to the *Tractatus*:

> Thus the aim of the book is to set a limit to thought, or rather – not to thought, but to the expression of thoughts: for in order to be able to set a limit to thought, we should have to find both sides of the limit thinkable (i.e. we should have to be able to think what cannot be thought).
>
> It will therefore only be in language that the limit can be set, and what lies on the other side of the limit will simply be nonsense. (p. 3)

Wittgenstein gives an interesting example of this theory in the following passage:

> 6.521   The solution of the problem of life is seen in the vanishing of the problem.
>
> (Is not this the reason why those who have found after a long period of doubt that the sense of life became clear to them have then been unable to say what constituted that sense?)
> 6.522   There are, indeed, things that cannot be put into words. They *make themselves manifest*. They are what is mystical.

Thus we might understand the meaning of life in a mystical experience, but we could not communicate this understanding in words. Similarly, it might become manifest to us that certain ethical claims were correct. But we would not then be able to put these claims into words.

Wittgenstein's argument in the *Tractatus* that metaphysical and, hence, most religious statements are meaningless might well seem to be an attack on religion; and indeed, it was interpreted as such by Carnap and other members of the Vienna Circle. In reality, however, Wittgenstein was offering an intellectually sophisticated defence of religion. We can appreciate the nature of this defence better by comparing it with the defence of religion offered by the Catholic Church. This is quite relevant to Wittgenstein, for, although his family were of Jewish origin, he was brought up as a Catholic.

Pope Leo XIII's encyclical *Aeterni Patris* of 1878 insisted that Catholicism should be based on scholastic philosophy, particularly on the writings of St Thomas Aquinas. Aquinas's philosophy is a metaphysical system which is essentially a modification of Aristotle's. So here religion is fitted into a metaphysical theory which is, naturally, accepted as meaningful. The trouble with grounding religious belief in this manner is that Aristotle's metaphysics has been somewhat discredited by the advance of science. As Martin aptly puts it: 'Aquinas's work was . . . embedded in an obsolete natural philosophy, of generally Aristotelian character, so that if his theology was recoverable, it had in some way to be reconciled with later scientific ideas' (1991, p. 43).

To put the problem in more general terms, if a particular religion is based on some metaphysical system, then the advance of science may undermine the credibility of the metaphysical theory adopted, and so undermine the religion itself. Suppose, however, following Wittgenstein, that we regard the doctrines of the religion as strictly meaningless and only to be apprehended by a mystical experience, and hold further that these religious doctrines are clearly demarcated from the meaningful statements of science. No advances in science will now have the effect of undermining the religion. Its doctrines are, so to speak, protected by being removed from the sphere of scientific criticism.

We have already remarked that Wittgenstein had discussions with Schlick and Waismann from 1927. Carnap attended some of these meetings, and has given an interesting account of them in his *Intellectual Autobiography*. He writes:

Once when Wittgenstein talked about religion, the contrast between his and Schlick's position became strikingly apparent. Both agreed of course in the view that the doctrines of religion in their various forms had no theoretical content. But Wittgenstein rejected Schlick's view that religion belonged to the childhood phase of humanity and would slowly disappear in the course of cultural development. . . .

These and similar occurrences in our conversations showed that there was a strong inner conflict in Wittgenstein between his emotional life and his intellectual thinking. His intellect, working with great intensity and penetrating power, had recognized that many statements in the field of religion and metaphysics did not, strictly speaking, say anything. In his characteristic absolute honesty with himself, he did not try to shut his eyes to this insight. But this result was extremely painful for him emotionally, as if he were compelled to admit a weakness in a beloved person. Schlick, and I, by contrast, had no love for metaphysics or metaphysical theology, and therefore could abandon them without inner conflict or regret. Earlier, when we were reading Wittgenstein's book in the Circle, I had erroneously believed that his attitude toward metaphysics was similar to ours. I had not paid sufficient attention to the statements in his book about the mystical, because his feelings and thoughts in this area were too divergent from mine. Only personal contact with him helped me to see more clearly his attitude at this point. I had the impression that his ambivalence with respect to metaphysics was only a special aspect of a more basic internal conflict in his personality from which he suffered deeply and painfully. (1963, pp. 26–7)

Carnap brings our clearly the difference between Wittgenstein's views and his own; but he tends to dismiss Wittgenstein's opinion as merely the result of 'a strong inner conflict', perhaps failing to see that it constitutes a subtle and sophisticated defence of religion.

Carnap enjoyed these discussions with Wittgenstein; but Wittgenstein did not reciprocate Carnap's friendly feelings, and, after a while, asked that Carnap should no longer attend his meetings with Schlick and Waismann. This is how Carnap recounts the matter:

When Wittgenstein talked about philosophical problems, about knowledge, language and the world, I usually was in agreement with his views and certainly his remarks were always illuminating and stimulating. Even at the times when the contrast in *Weltanschauung* and basic personal attitude became apparent, I found the association with him most interesting, exciting and rewarding. Therefore I regretted it when he broke off the contact. From the beginning of 1929 on, Wittgenstein wished to meet only with Schlick and

Waismann, no longer with me or Feigl, who had also become acquainted with him in the meantime, let alone with the Circle. Although the difference in our attitudes and personalities expressed itself only on certain occasions, I understood very well that Wittgenstein felt it all the time and, unlike me, was disturbed by it. He said to Schlick that he could talk only with somebody who 'holds his hand'. (p. 27)

Wittgenstein's theory of philosophy poses a problem for his own book, the *Tractatus*. In the *Tractatus* Wittgenstein claims that philosophical writings are meaningless, but, since the *Tractatus* is itself a piece of philosophical writing, it would seem to follow that the *Tractatus* itself must be meaningless. With a surprising (though not uncharacteristic) rigour, Wittgenstein accepts this conclusion. The doctrines of the *Tractatus* 'make themselves manifest' as true; but they cannot be meaningfully stated in words. This position is put forward in the famous penultimate proposition of the *Tractatus*.

6.54 My propositions serve as elucidations in the following way: anyone who understands me eventually recognises them as non-sensical, when he has used them – as steps – to climb up beyond them. (He must, so to speak, throw away the ladder after he has climbed up it.)

He must transcend these propositions, and then he will see the world aright.

This proposition of Wittgenstein's has undoubtedly a certain dramatic appeal. Yet, we should really feel considerable disquiet when an author states on the last page that his book is meaningless. Surely something has gone wrong! Something had indeed gone wrong, and this was demonstrated by Popper in his critique of the Vienna Circle, which we shall consider in the next section.

8.5 Popper's Critique of the Vienna Circle on Metaphysics

Popper put forward two basic criticisms of the Vienna Circle's views on science and metaphysics. First of all, he proposed that *verifiability* should be replaced by *falsifiability* as the criterion of demarcation between science and metaphysics. Secondly, he claimed that metaphysics, though different from science, was in general meaningful, and might even be positively helpful to science in some cases. For Popper, the demarcation between science and metaphysics is not a demarcation between sense and nonsense.

It seems to me very important to distinguish carefully between these two criticisms, for the following reason. The falsifiability criterion has been subjected to a number of quite technical criticisms, but even if some of these are correct, it by no means follows that we should reject Popper's views on science and metaphysics *in toto*. As a matter of fact, Popper's second thesis, that metaphysics is, in general, meaningful, and may, in some cases, be positively helpful to science, is largely independent of the exact details of the demarcation criterion. It is merely necessary that some, at least rough, demarcation between science and metaphysics should be possible. Moreover, this thesis of Popper's is of great importance, since, somewhat surprisingly, even today, many philosophical schools dismiss metaphysics as meaningless or, at least, undesirable. Yet Popper's general arguments in favour of metaphysics are, as we shall see, almost wholly convincing.

Popper first published an extended account of his criticisms of the Vienna Circle in his *Logic of Scientific Discovery* of 1934. Another useful text is chapter 11 of his *Conjectures and Refutations* of 1963. This chapter is entitled 'The Demarcation between Science and Metaphysics', and was written in 1955 as Popper's contribution to the volume on the philosophy of Rudolf Carnap in the Library of Living Philosophers series edited by P. A. Schilpp. So in some parts of this chapter, Popper criticizes the views of Carnap which we expounded in the previous section. In particular, Popper formulates his criticism of verifiability as a demarcation criterion as follows: 'My criticism of the verifiability criterion has always been this: against the intention of its defenders, *it did not exclude obvious metaphysical statements; but it did exclude the most important and interesting of all scientific statements*, that is to say, the scientific theories, *the universal laws* of nature' (1963, p. 281). Let us take the second part of this first. As usual, it is easiest to start with the simple philosophers' example of a universal generalization: namely, 'All ravens are black'. This is not verifiable by any finite conjunction of observation statements about ravens; but it is falsifiable by observing a white raven. Indeed, the analogous generalization 'All swans are white' was falsified in just this way. There is thus a *logical asymmetry* between verifiability and falsifiability, as regards such universal generalizations. As Popper puts it: 'My proposal is based upon an *asymmetry* between verifiability and falsifiability; an asymmetry which results from the logical form of universal statements. For these are never derivable from singular statements, but can be contradicted by singular statements' (1934, p. 41).

This point can be amplified by introducing the consideration of *existential statements*. An existential statement is one which asserts that something exists. The statement 'There is (or there exists) a white raven' is a simple example of an existential statement. Now the interesting thing is that the situation regarding the verifiability and falsifiability of existential statements such as 'There is a white raven' is exactly the opposite of what obtains for universal statements such as 'All ravens are black'. As we have seen, the universal statement 'All ravens are black' is not verifiable by an observation statement, but is falsifiable by such a statement – namely, one reporting the observation of a non-black raven. The existential statement 'There is a white raven', on the other hand, is verifiable by an observation statement – namely, one which reports the observation of a white raven – but it is not falsifiable by an observation statement. However many ravens we observe and whatever their colours, our observations cannot possibly contradict the statement that there is a white raven.

This remark about existential statements brings us to the other half of Popper's criticism of the verifiability criterion: namely, his claim that '*it did not exclude obvious metaphysical statements*' (1963, p. 281). What Popper has in mind here are, *inter alia*, existential statements drawn from religion, magic, and the occult. An example would be 'The Devil exists', or, as Popper puts it with more precision and detail:

> My example consists of the following purely existential theory:
>   'There exists a finite sequence of Latin elegiac couplets such that, if it is pronounced in an appropriate manner at a certain time and place, this is immediately followed by the appearance of the Devil – that is to say, of a man-like creature with two small horns and one cloven hoof.'
>   Clearly, this untestable theory is, in principle, verifiable. (1963, p. 249)

This incantation statement of Popper's (as it might be called) is indeed verifiable; but presumably it is the kind of occult claim that we would wish to exclude from science. Another example, along the same lines, is 'There exists conscious experience after death'. It is logically possible for this statement to be verified, but it cannot be falsified. Again, this is a statement which belongs more to religion than to science.

After these preliminaries, let us now turn to Popper's formulation of falsifiability as a demarcation criterion. This is how he puts the matter:

> Theories are . . . *never* empirically verifiable. If we wish to avoid the positivist's mistake of eliminating, by our criterion of demarcation, the theoretical systems of natural science, then we must choose a criterion which allows us to admit to the domain of empirical science even statements which cannot be verified.
>
> But I shall certainly admit a system as empirical or scientific only if it is capable of being *tested* by experience. These considerations suggest that not the *verifiability* but the *falsifiability* of a system is to be taken as a criterion of demarcation. (1934, p. 40)

Note that Popper does not here speak of a theory, but of 'a system'. This is an important point, to which we shall return later. Now, however, I will turn to Popper's second, more general, criticism of the Vienna Circle.

The Vienna Circle regarded verifiability as a *meaning criterion*. Popper, by contrast, put forward falsifiability as a demarcation, but *not* a meaning, criterion. Indeed, Popper always maintained that many metaphysical statements are perfectly meaningful.

Popper has a simple, but powerful, argument to show that neither verifiability nor falsifiability are adequate as meaning criteria (cf. 1983, pp. 177–8). It depends on the principle that if a statement, S, is meaningful, then its negation, not-S, must be meaningful. If we take S to be a universal generalisation, we get the interesting result that its negation, not-S, is an existential statement. More specifically we have the following:

S = All ravens are black
not-S = It is not the case that all ravens are black
= There is (or exists) a non-black raven

Now, as we have already pointed out, S is falsifiable, but not-S is not falsifiable. Thus, if we were to adopted falsifiability as a meaning criterion, we would have to say that S is meaningful and that not-S is meaningless; but this, to say the least, is highly counter-intuitive. Exactly the same argument can be used against verifiability as a meaning criterion, since not-S is verifiable, while its negation, not-not-S = S, is falsifiable but not verifiable.

In the next chapter we shall consider Popper's view that meta-physical ideas and theories can be – and indeed have been – useful for science. This consideration provides further arguments for the claim that metaphysical speculations are often meaningful, since it seems unlikely that such speculations could be helpful for science if they were meaningless or mere nonsense. Before coming to this, however, I will present in the next section an argument of my own for the thesis that much metaphysics is in fact meaningful. Curiously enough, this argument is based on some ideas of Popper's principal opponent, Wittgenstein.

At first sight, nothing could seem more paradoxical than to use some of Wittgenstein's ideas in support of Popper. Popper, as we have seen, criticized Wittgenstein's *Tractatus* in a decisive way. The famous meeting of Popper and Wittgenstein at Cambridge (described in 1.5) has an almost mythic quality; but, whatever exactly happened, it is clear that the two famous philosophers did not part on friendly terms. Now, admittedly, Wittgenstein himself criticized his earlier *Tractatus* views, and developed a quite different approach in his later work, the *Philosophical Investigations*. Popper, however, was, if anything, even less appreciative of this later work than he had been of the *Tractatus*. Indeed, he went so far as to say in a radio programme broadcast in the winter of 1970–1: 'Russell read the *Philosophical Investigations* without deriving any enlightenment from it. So did I, I must admit' (Magee (ed.), 1971, p. 138). Yet, when the dust settles on a heated philosophical controversy, it is sometimes possible to see that the embattled opponents had more in common than they themselves realized at the time. In his *Philosophical Investigations* Wittgenstein develops a new theory of meaning. I will briefly expound this theory, and then argue that it can be used to defend Popper's thesis that many metaphysical theories are meaningful.

8.6   Wittgenstein's Later Theory of Meaning

Wittgenstein's theory in his *Philosophical Investigations* is that the meaning of a word is given by its use in a language-game. By a 'language-game' he means some kind of rule-guided social activity in which the use of language plays an essential part. He himself introduces the concept as follows: 'I shall also call the whole, consisting of language and the actions into which it is woven, the "language-game"' (1953, sec. 7). And again: 'Here the term "language-*game*" is meant to bring into prominence the fact that the *speaking* of language is part of an activity, or of a form of life' (sec. 23).

Wittgenstein illustrates his concept of language-game by his famous example involving a boss and a worker on a building site. The boss shouts 'Slab', for example, and the worker has to fetch a slab. Wittgenstein's point is that the meaning of the word 'slab' is given by its use in the activity carried out by boss and worker.

I am broadly in sympathy with Wittgenstein's *Investigations* theory of language; but his term 'language-*game*' seems to me inappropriate. In fact, his first example of a language 'game' is not a game at all, but work. I prefer therefore to speak in general of 'language-activities', reserving the term 'language-game' for those activities which really are games. Interestingly enough, the 'meaning = use' theory is partly anticipated by Wittgenstein in the *Tractatus* where he writes:

> 6.211   (In philosophy the question, 'What do we actually use this word or this proposition for?' repeatedly leads to valuable insights.)

Let us now see how Wittgenstein's 'meaning = use' theory can be used to defend the Popperian thesis that metaphysical statements are in general meaningful. Imagine a group of people who meet regularly (on Wednesday afternoons, say) to discuss metaphysical questions. They could be Catholic theologians, for example, or Hegelian philosophers. They will certainly use a large number of curious words and expressions, such as 'essence', 'ground of being', 'dialectic', and so on. Yet, this discourse is not arbitrary, but rule-guided. A beginner who uses an expression incorrectly is reprimanded, and may even be ostracized if he or she does not conform. Within the group it is well known who are the experts whose pronouncements are listened to with most respect, and so on. Here, surely, we have a language-game (the term 'game' is perhaps more appropriate in this instance), a rule-guided activity, or a form of life. Within this language-game, words and expressions have a use which is circumscribed by rules and conventions. On Wittgenstein's later theory of meaning, therefore, we must surely say that these words and expressions have meaning, and that the metaphysical discourse is meaningful.

Would Wittgenstein himself have agreed with this use of his *Investigations* theory of meaning to support his rival Popper? There is one passage which suggests that he might have done. In section 23 of the *Philosophical Investigations* he reviews the multiplicity of language-games by giving a long list of examples. This list includes 'praying'. This suggests that Wittgenstein would have included religious ceremonies as language-games, and so would have been committed to the view that religious discourse in general, and theology in particular, was meaningful.

There are a few indications (particularly the passage just cited) that Wittgenstein moved, in his later period, towards the view that religious and metaphysical discourse were meaningful. But these indications are few, and my overall impression is that Wittgenstein in the *Philosophical Investigations* by and large stuck to the *Tractatus* view that metaphysical statements are meaningless. Thus he says in section 464: 'My aim is: to teach you to pass from a piece of disguised nonsense to something that is patent nonsense.' This is very similar to the passage from 6.53 of the *Tractatus* cited earlier, namely: 'The correct method in philosophy would really be the following:... whenever someone else wanted to say something metaphysical, to demonstrate to him that he had failed to give a meaning to certain signs in his propositions.'

This situation leads to what I regard as a central flaw in Wittgenstein's later philosophy. The view that metaphysics is meaningless does indeed follow from the *Tractatus* theory of meaning, but that theory of meaning is inadequate. In the *Philosophical Investigations*, Wittgenstein puts forward a much better theory of meaning, but makes the mistake of holding on to his old view of the senselessness of metaphysics – a view which is actually incompatible with his new account of meaning.

Such, at all events, is my own opinion; but some defenders of Wittgenstein might still think that the nonsensical character of metaphysics could be defended in terms of the *Investigations* theory of meaning, despite the considerations given earlier. Let us next examine how this could be done.

Now it will be remembered that Wittgenstein's first example of a language activity was work on a building site. The boss shouts 'Slab', and a worker goes off and brings a slab. This rule-guided social activity of boss and worker gives meaning to the word 'slab'. Let us contrast this with the metaphysical discussion group which meets on Wednesday afternoons and considers such questions as whether the essence of the ground of being implies existence, and so on. Perhaps Wittgenstein might say that words acquire meaning in the practical everyday social activities of the building site, but not in the purely theoretical discussions of groups of philosophers. There are indeed hints of such a view in his writing – in the following passage for example:

We are under the illusion that what is peculiar, profound, essential, in our investigation, resides in its trying to grasp the incomparable essence of language. That is, the order existing between the concepts of proposition, word, proof, truth, experience, and so on. This order is a *super*-order between – so to speak – *super*-concepts. Whereas, of course, if the words 'language', 'experience', 'world', have a use, it must be as humble a one as that of the words 'table', 'lamp', 'door'. (1953, sec. 97)

It looks here as if a genuine use of a word can only be a 'humble' one.

The view we are considering has a certain populist flavour. Words acquire genuine meanings in practical, everyday social activities in which workers heave slabs across building sites. The purely theoretical language-games of (effete?) intellectuals do not suffice to give their abstract terms meaning. As we have already remarked, Wittgenstein seems to have been attracted emotionally towards a populism of this kind. He admired Tolstoy's eulogies of virtuous peasants, and disliked the conversation of his fellow academics. The following passage from Karl Britton's 'Portrait of a Philosopher' shows a personal attitude not unrelated to the philosophical view we are now considering:

He had, he said, only once been to high table at Trinity and the clever conversation of the dons had so horrified him that he had come out with both hands over his ears. The dons talked like that only to score: they did not even enjoy doing it. He said his own bedmaker's conversation, about the private lives of her previous gentlemen and about her own family, was far preferable: at least he could understand why she talked that way and could believe that she enjoyed it. (Quoted from Pitcher, 1964, p. 12)

Whatever sympathies we may have with the populist contrast between workers, who are 'real people', as against dons, who are 'atrophied intellectuals', the theory we are considering is untenable, as the following counter-example shows. Let us suppose that a piece of pure mathematics is developed by a group of pure mathematicians. It is then taken up by a group of theoretical physicists, and used in the creation of a new physical theory. Finally, this theory is used in a practical application – perhaps even in house building. Suppose, further, that we adopt the view that a term is meaningful only if it is used in a practical, everyday social activity and not when it is used in purely theoretical discourse. It then follows that the terms of the mathematical theory are meaningless while the theory is being developed by the pure mathematicians, that they remain meaningless when the mathematics is used to create the new physical theory, but then suddenly become meaningful for the first time when that theory is applied to house building. Such a consequence seems to me quite unacceptable.

Our hypothetical example is, of course, no mere fantasy, but is quite close to some actual historical cases. For example, tensor analysis was developed as a piece of pure mathematics by two Italians, Ricci and Levi-Cività, in an important paper of 1901. It was then used by Einstein in his general theory of relativity of 1915. The general theory of relativity has recently been applied to practical problems concerned with satellites. At what stage did the key terms of the tensor calculus becomes meaningful? It seems clear that they were meaningful throughout the process.

We cannot, therefore, use the demarcation between practical discourse and purely theoretical discourse to distinguish sense from nonsense. Yet the practical/purely theoretical demarcation is, like the science/metaphysics demarcation, of great interest. The two distinctions are not, however, the same, as I will now demonstrate. Let us begin with an example of a scientific theory which is not used in practice, and then go on to an example of a theory which is used in practice but is not scientific. Einstein's general theory of relativity was introduced in 1915, but has been used in practical applications only recently. For more than fifty years, therefore, this theory was a scientific theory, but one not used in practice. Conversely, it is very easy to find examples of theories which are not scientific, but are used in practice. Consider a tribe which has a mythological theory of the elements, and uses this theory in its rain-making dances and other agricultural practices. Here we have an example of a theory which is unscientific, but which is used in practice. As this may seem to exhibit a rather patronizing attitude towards primitive peoples, it should be pointed out that there may well be similar examples in our own society. Politicians regularly use curious economic theories in deciding government policy, and it is often by no means clear that these economic theories are scientific.

8.7 The Influence of Wittgenstein's Life on his Philosophy

Having now considered some of Wittgenstein's philosophical views, let us return to the interesting question of how these are related to his strange life. The most striking thing about Wittgenstein as a philosopher is that he produced two quite distinct philosophies. What led him to abandon his views in the *Tractatus*? And what provided the stimulus to produce the new philosophy to be found in his *Philosophical Investigations*? Von Wright has this to say in answer to these questions:

> Of great importance in the origination of Wittgenstein's new ideas was the criticism to which his earlier views were subjected by two of his friends. One was Ramsey, whose premature death in 1930 was a heavy loss to contemporary thought. The other was Piero Sraffa, an Italian economist who had come to Cambridge shortly before Wittgenstein returned there. It was above all Sraffa's acute and forceful criticism that compelled Wittgenstein to abandon his earlier views and set out upon new roads. He said that his discussions with Sraffa made him feel like a tree from which all branches had been cut. That this tree could become green again was due to its own vitality. The later Wittgenstein did not receive an inspiration from outside like that which the earlier Wittgenstein got from Frege and Russell. (1958, p. 5)

Now, without wishing to deny that the criticisms of Ramsey and Sraffa were of some importance, I would like to suggest that the principal stimulus for changing his philosophy came from Wittgenstein's life experiences between 1920 and 1929, and that these experiences were the sources of the vitality which enabled the Wittgensteinian tree to become green again.

> Grau, teurer Freund, is alle Theorie
> Und grün des Lebens goldner Baum.
> Goethe, *Faust, Part I*

(All theory, dear friend, is grey, and green the golden tree of life.)

When Wittgenstein began writing the *Tractatus*, his previous training had been in the physical sciences, engineering, and logic; and his views about the nature of language reflect this background. Language consists of elementary propositions which picture reality and which are glued together by logical connectives to form complex propositions. Meaningful propositions are either logical contradictions or logically valid propositions or empirical propositions of the natural sciences. A model for all this is provided by Newtonian mechanics, for Wittgenstein himself says: '6.341

Newtonian mechanics, for example, imposes a unified form on the description of the world.'

It is not surprising that the Vienna Circle, absorbed as its members were in the study of logic, mathematics, and science, should find this attractive and plausible. But in the 1920s the author of the *Tractatus* was involved in trying to teach language to young children. Indeed, Wittgenstein's next publication after the *Tractatus* was his *Wörterbuch für Volksschulen* (1926). Despite the unfortunate end to the author's career as a teacher, this was officially approved as a school textbook. (For more details about the book and how it was compiled, see Bartley, 1973, p. 77.) Now Wittgenstein could hardly have failed to notice that the *Tractatus* theory of language was not very plausible as an account of how his young pupils used language. Moreover, there are signs everywhere in the *Philosophical Investigations* of the influence of Wittgenstein's years as a schoolmaster. Thus, already in section 9, when Wittgenstein begins a discussion of the natural numbers, he remarks: 'Children do learn the use of the first five or six cardinal numerals in this way.' Sections 156–78 are devoted to a discussion of reading; while in section 185, where Wittgenstein introduces what have come to be known as the rule-following considerations, we find the following:

> Now – judged by the usual criteria – the pupil has mastered the series of natural numbers. Next we teach him to write down other series of cardinal numbers and get him to the point of writing down series of the form

> 0, n, 2n, 3n, etc.

> at an order of the form '+n'; so at the order '+1' he writes down the series of natural numbers. – Let us suppose we have done exercises and given him tests up to 1000.
> Now we get the pupil to continue a series (say +2) beyond 1000 – and he writes 1000, 1004, 1008, 1012.
> We say to him: 'Look what you've done!' – He doesn't understand. We say: 'You were meant to add *two*: look how you began the series!' – He answers: 'Yes, isn't it right? I thought that was how I was *meant* to do it.'

The example is, of course, fanciful, but it does not sound too far removed from Wittgenstein's own classroom experience.

The influence of Wittgenstein's work on his sister's mansion is even more obvious, since, as we have seen, the very example with which Wittgenstein introduces his concept of 'language-game' is that of a boss and a worker on a building site. Thus, although Wittgenstein appears to have abandoned philosophy for a while in order to do other things, these other things were actually a kind of preparation for his later innovations in philosophy.

9

Metaphysics in Relation to Science: The Views of Popper, Duhem, and Quine

9.1 Popper on Metaphysics in Relation to Science

One of the most important aspects of Popper's critique of the Vienna Circle is his insistence that metaphysics can be not only meaningful, but even of positive value to science. Popper holds that theories may start life as metaphysical, but then come gradually to be transformed into scientific hypotheses. In the following well-known passage, he compares this process to that of particles in a liquid being gradually deposited at the bottom of the container:

To obtain a picture or model of this quasi-inductive evolution of science, the various ideas and hypotheses might be visualized as particles suspended in a fluid. Testable science is the precipitation of these particles at the bottom of the vessel: they settle down in layers (of universality). The thickness of the deposit grows with the number of these layers, every new layer corresponding to a theory more universal than those beneath it. As the result of this process ideas previously floating in higher metaphysical regions may sometimes be reached by the growth of science, and thus make contact with it, and settle. Examples of such ideas are atomism; the idea of a single physical 'principle' or ultimate element (from which the others derive); the theory of terrestrial motion (opposed by Bacon as fictitious); the age-old corpuscular theory of light; the fluid-theory of electricity (revived as the electron-gas hypothesis of metallic conduction). All these metaphysical concepts and ideas may have helped, even in their early forms, to bring order into man's picture of the world, and in some cases they may even have led to successful predictions. Yet an idea of this kind acquires scientific status only when it is presented in falsifiable form; that is to say, only when it becomes possible to decide empirically between it and some rival theory. (1934, pp. 277–8)

Of the various examples which Popper cites in this passage, the most striking, perhaps, is that of *atomism*. Atomism was first introduced in the West by the pre-Socratic thinkers Leucippus and Democritus. It continued as a powerful trend in the ancient world with Epicurus in Greece and Lucretius in Rome. This ancient atomism must, I think, be classed as metaphysical rather than scientific.

Ancient atomism was revived in Western Europe in the seventeenth century, and discussed by the leading scientists of the day. It should still, at that time, be considered as a metaphysical rather than a scientific hypothesis. At the beginning of the nineteenth century, Dalton reintroduced atomism to solve some problems in chemistry; while half-way through the nineteenth century, Maxwell brought atomism into mathematical physics in connection with the kinetic theory of gases. By the end of the nineteenth century, atomism can definitely be considered as a scientific hypothesis; but this scientific development would scarcely have been possible without the earlier history of atomism as metaphysics.

This example constitutes yet another argument for the thesis that metaphysical theories can be perfectly meaningful. Anyone who maintains that metaphysics is always meaningless is faced with some difficult choices. One option would be to say that atomism was always scientific, and so meaningful; but then it has to be maintained that the atomism of the ancient Greeks was scientific, and this is hardly plausible. The other option would be to say that atomism was meaningless from ancient Greek times until some point in the nineteenth century, when it became scientific and so meaningful; but then it must be held that a perfectly meaningless theory was somehow used by Dalton, Maxwell, and others in order to formulate their scientific and meaningful theories. This again is an absurd and untenable position. Thus there is no escape from the conclusion that metaphysics can be both meaningful and useful for scientific progress.

In his *Realism and the Aim of Science* of 1983, Popper develops his view on metaphysics by introducing the concept of a *metaphysical research programme* for science. Thus, he says: 'Atomism is an excellent example of a non-testable metaphysical theory whose influence upon science has exceeded that of many testable scientific theories' (1983, p. 192). After giving some further example of metaphysical theories which have influenced science, he continues:

'Each of these metaphysical theories served, before it became testable, as a research programme for science. It indicated the direction of our search, and the kind of explanation that might satisfy us; and it made possible something like an appraisal of the depth of a theory' (pp. 192–3). This is an important passage, because it indicates the *heuristic* role of metaphysics in guiding the construction of scientific hypotheses. Such hypotheses almost always emerge in the course of a scientist or a group of scientists pursuing a research programme, and such research programmes are usually guided by some general (or metaphysical) principles and ideas which indicate the kind of specific hypothesis which should be devised to explain existing facts and to be tested against further observations and experiments. Thus the general ideas of atomism guided Dalton in constructing a hypothesis to explain some facts of chemical combination, and guided Maxwell in trying to account for the observed relations of pressure, volume, and temperature in gases. Without the metaphysical ideas of atomism to guide their research programmes, it is very doubtful whether Dalton or Maxwell could have devised their specific scientific hypotheses.

Popper was undoubtedly right to rehabilitate metaphysics against the strictures of the Vienna Circle, but it still seems to me that some restraints should be put on enthusiasm for metaphysics. Many metaphysical theories – for example, Leibniz's Monadology – have not affected science very much one way or the other, while some metaphysical theories have perhaps impeded, rather than helped, science. For example, religious theories of divine creation and the human soul were (and perhaps still are, to some extent) obstacles to the Darwinian theory of biological evolution.

So far we have discussed the influence of metaphysics on science; but, conversely, science can influence metaphysics. I will conclude this section with a brief discussion of this opposite influence. An obvious example here is the debate between determinism and indeterminism. Laplace adopted a strong version of determinism, and was clearly influenced by the success of Newtonian mechanics. Conversely, the discovery of limitations on Newtonian mechanics and the adoption, as a fundamental theory in physics, of quantum mechanics, with its probability and uncertainty, has led to a revival of indeterminism. Indeed, Popper has written a book, *The Open Universe* (1982), arguing the case for indeterminism.

Some possible misconceptions on this matter must, however, be avoided. I am not maintaining – and it is not the case – that a new scientific theory logically implies a new metaphysics. Popper, for example, argues convincingly in the book just cited that Newtonian mechanics, even if accepted as the fundamental theory of physics, does not necessitate the adoption of metaphysical determinism. Yet there is no doubt that Laplace was influenced by the success of Newtonian mechanics in formulating his views on determinism. Conversely, Popper does not argue from quantum mechanics to metaphysical indeterminism. Indeed he writes: 'My refutation of "scientific" determinism . . . nowhere makes use of probability theory; nor do I appeal to quantum theory. "Free will", too, is mentioned only incidentally. . . . My argument holds for every physical theory, however strongly deterministic it may appear' (1982, p. 106). Yet Popper, as well as other recent advocates of indeterminism, are undoubtedly influenced by the success of quantum mechanics.

9.2   Duhem and Quine on the Status of Metaphysics

From the very beginning of his classic *The Aim and Structure of Physical Theory*, Duhem presupposes that metaphysical systems exist and that these systems, though distinct from those of science, are nonetheless meaningful. I say advisedly that Duhem *presupposes* these doctrines, since he nowhere states them, but rather assumes them when arguing for some other thesis. Thus part 1, chapter 1, of *The Aim and Structure* is entitled 'Physical Theory and Metaphysical Explanation'. Here Duhem introduces metaphysics in explicit contrast to physics, in the following passage:

> Now these two questions – Does there exist a material reality distinct from sensible appearances? and What is the nature of this reality? – do not have their source in experimental method, which is acquainted only with sensible appearances and can discover nothing beyond them. The resolution of these questions transcends the methods used by physics; it is the object of metaphysics. (1904–5, p. 10)

Duhem goes on to argue that theoretical physics should not be subordinated to metaphysics. His reason is the following: 'If theoretical physics is subordinated to metaphysics, the divisions separating the diverse metaphysical systems will extend into the domain of physics. A physical theory reputed to be satisfactory by the sectarians of one metaphysical school will be rejected by the partisans of another school' (pp. 10–11). He then goes on to consider an example: namely, the theory of the action of a magnet on iron. With a characteristic wealth of historical detail, he shows how attempts were made to subordinate this physical theory to different systems of metaphysics, such as Aristotelianism, Boscovich's point-atomism, material atomism, and Cartesianism. These attempts were all, so Duhem argues, unsatisfactory.

I do not want to consider here whether Duhem is right or not in making this particular claim. My point is only that he assumes, throughout the discussion, the existence of meaningful metaphysical systems such as Aristotelianism, atomism, and Cartesianism, which are distinct from the scientific systems of physics. Today it may seem surprising to make assumptions of this kind without any discussion or comment, but it must be remembered that Duhem was writing before Wittgenstein and the Vienna Circle launched their attack on metaphysics. Duhem was a man well versed in the history of both philosophy and science, and was familiar, therefore, with a considerable number of different metaphysical systems which would have appeared to him as obviously both meaningful and distinct from science. Doubts about metaphysics arose later from the logical analysis and theories of meaning of Russell, Wittgenstein, and the Vienna Circle.

For the same reason, Duhem does not attempt in his (1904–5) to formulate a criterion for demarcating science from metaphysics – though he assumes that the two are distinct. The problem of finding an adequate demarcation criterion in terms of some concept such as verifiability or falsifiability arose out of the attempt to analyze the structure of scientific theories with the tools of formal logic. Duhem reasons very logically, but he uses logic in an informal fashion. He nowhere employs the formal logic of Frege, Peano, and Russell (see chapter 3, note 4).

Let us next compare Duhem's views on science and metaphysics with those of Quine. Once again it will become clear that the views of the two thinkers are quite distinct. Vuillemin puts the point as follows: 'The D-thesis . . . arose from historical considerations and only physics fell under its scope, metaphysics and science being kept completely distinct, Quine . . . refrained from drawing any sharp distinction between science and metaphysics' (1979, p. 609). This is certainly correct. Duhem, as we have seen, clearly assumes that science and metaphysics are distinct, while Quine has this to say: 'Both dogmas, I shall argue, are ill-founded. One effect of abandoning them is, as we shall see, a blurring of the supposed boundary between speculative metaphysics and natural science' (1951, p. 20).

On this issue, my own sympathies are with Duhem rather than Quine. Typical metaphysical theories such as Plato's theory of Forms or the doctrine of the Trinity do seem to have a different character from typical scientific theories such as Lavoisier's theory of combustion or Bohr's theory of the atom. The boundary between metaphysics and science is thus real, rather than 'supposed'. Of course, the boundary is not entirely sharp, and intermediate or doubtful cases do undoubtedly occur; but this is true of many distinctions which are nonetheless perfectly viable and useful. In everyday life we can easily enough distinguish between those who are bald and those who are not, even though it is hard to say how many hairs a man must lose before he becomes bald!

Another consideration relevant to this subject is the following. Some of Quine's modes of expression are arguably quite misleading. Thus Quine writes: 'The totality of our so-called knowledge or beliefs, from the most casual matters of geography and history to the profoundest laws of atomic physics or even pure mathematics and logic, is a man-made fabric which impinges on experience only along the edges' (1951, p. 42). Here Quine presupposes that he, his readers, and perhaps a wider circle of intellectuals share a totality of *our* beliefs. But is it in fact correct to speak of *our* beliefs when beliefs vary so widely from one person to another? Consider, for example, a group of theoretical physicists. This group might contain a devout Roman Catholic, an orthodox Jew, a Bible-belt Protestant, a religiously agnostic admirer of the free market, a liberal Muslim, a Marxist-Leninist, and so on. Anyone familiar with the theoretical physics community will recognise that this example is quite realistic. The point, however, is that the various beliefs of the group do not form a totality shared by the group as a whole. There will, of course, be a core of physical theories which nearly everyone in the group will agree on accepting at least provisionally, but there will also be a host of religious and metaphysical beliefs about which different members of the group will disagree most strongly. These religious and metaphysical differences may have effects even within physics itself, and will very likely, as we have already argued, influence the way in which the various physicists carry out their research. However, the general situation is almost impossible to describe without making the science/metaphysics distinction, and Quine's picture of knowledge or beliefs as forming a seamless garment does not accord at all well with the realities of the situation.

9.3   Duhem and Popper on the Influence of Metaphysics on Science

Duhem's views on metaphysics and science are in fact much more like Popper's than Quine's. Duhem and Popper agree, in contrast to Quine, that science can be distinguished from metaphysics, though Duhem does not in his 1904–5 attempt to formulate a demarcation criterion. Duhem is perhaps a little less enthusiastic than Popper about metaphysics in relation to science. As we have seen, Duhem stresses that physics should not be subordinated to metaphysics. On the other hand, he does admit that metaphysics can influence science, and gives many historical examples, some of which overlap those considered by Popper. We will next consider a few of these Duhemian examples of metaphysical influence on science, as they are both interesting and instructive.

Let us start with Descartes' metaphysical views about matter, which, Duhem argues, later influenced the hydrodynamics of W. Thomson and Maxwell's theory of magnetism. This is how Duhem describes this example:

> According to Descartes, matter is essentially identical with the extended in length, breadth, and depth, as the language of geometry goes; we have to consider only its various shapes and motions. Matter for the Cartesians is, if you please, a kind of vast fluid, incompressible and absolutely homogeneous. Hard, unbreakable atoms and the empty spaces separating them are merely so many appearances, so many illusions. Certain parts of the universal fluid may be animated by constant whirling or vortical motions; to the coarse eyes of the atomist these whirlpools or vortices will look like individual corpuscles. The intermediary fluid transmits from one vortex to the other forces which Newtonians, through insufficient analysis, will take for actions at a distance. Such are the principles of the physics first sketched by Descartes, which Malebranche investigated further, and to which W. Thomson, aided by the hydrodynamic researches of Cauchy and Helmholtz, has given the elaboration and precision characteristic of present-day mathematical doctrines.
>
> This Cartesian physics cannot dispense with a theory of magnetism; Descartes had already tried to construct such a theory. The corkscrews of 'subtle matter' with which Descartes, not without some naïveté, in his theory replaced the magnetic corpuscles of Gassendi were succeeded, among the Cartesians of the nineteenth century, by the vortices conceived more scientifically by Maxwell. (1904–5, p. 13)

Duhem also discusses at some length the surprising fact that unscientific astrology had, at one stage, a beneficial influence on mainstream science. Duhem introduces this theme thus: 'Discovery is not subject to any fixed rule. There is no doctrine so foolish that it may not some day be able to give birth to a new and happy idea. Judicial astrology has played its part in the development of the principles of celestial mechanics' (p. 98). 'Judicial astrology', in fact, proved helpful in the development of the Newtonian theory of universal gravitation, and, in particular, of the theory that the tides are caused by the attraction of the Moon. Now astrology is based on the idea that the heavenly bodies influence the fate of mankind here on earth; so astrologers would naturally be sympathetic to the view that the tides are caused by the Moon's attraction, since this would appear to them as just one instance of their general idea that the heavens influence events here on earth. Conversely, many of the more standard schools of science rejected the doctrine of the Moon's attraction as occult and superstitious. So, in this instance, the ideas of the less rational astrologers proved more successful than those of their more rational contemporaries. This is how Duhem describes the development:

> Ptolemy and Albumasar did not hesitate to invoke a particular virtue, a special influence of the moon on the waters of the sea. Such an explanation was not intended to please the true disciples of Aristotle; . . .
> The virtue that the tides manifest was, on the other hand, made to order for the astrologers who found in it the undeniable proof of the influences that the heavenly bodies exert on sublunar things. This hypothesis was in no less favour among the physicians who compared the role played by heavenly bodies in the tidal phenomenon with the role attributed to them in crises of disease; did not Galen attach the 'critical days of pituitary diseases' to the phases of the moon? (pp. 233–4)

Interestingly enough, the astrologers did not just formulate the main theory (that the Moon's attraction is the cause of the tides), but also recognized that variations in the tides are due to the influence of the Sun. As Duhem says:

> Morin fell back on principles of judiciary astrology in order to affirm the role played by the sun in the variations of tide, and it is indeed to the indisputable credit of the astrologers that they prepared all the materials for the Newtonian theory of tides, whereas the defenders of

rational scientific methods, Aristotelians, Copernicans, atomists, and Cartesians, have in emulation fought its advent. (p. 240)

It is somewhat misleading of Duhem to include Copernicans in this list. He is clearly thinking of Galileo, who denied that tides were caused by the Moon's attraction; but another famous Copernican, Kepler, held that the Moon attracts the waters of the sea by a magnetic action. Duhem's mention of the Cartesians in this context is, however, quite appropriate, and is worth pursuing.

We have seen that Descartes' metaphysical views of the nature of matter had a beneficial influence in the nineteenth century on the development of hydrodynamics and Maxwell's theory of magnetism. At an earlier stage, however, Cartesian metaphysics acted as a definite block to scientific progress, and was an obstacle to the acceptance of Newton's theory of gravity. It was part of Descartes' system that one piece of matter could affect another piece of matter only through direct contact. Action at a distance was not allowed. Yet Newton's law of gravity states that a piece of matter exerts a force on every other piece of matter, however distant, according to the inverse square relation. Such a law was clearly quite inadmissible according to Cartesian metaphysics. This situation led Newton himself to doubt his own law, and to hope that a deeper explanation which avoided the use of action at a distance would be found. Naturally, the reaction of many Cartesians to Newton's law was still more hostile.

There are undoubted examples of metaphysics exerting a beneficial influence on the progress of science, but we should not forget that metaphysical ideas can obstruct, as well as help, science. What is interesting about the present example of Cartesian metaphysics is that it shows that the very same metaphysical system can be helpful in one scientific context and obstructive in another.

In general terms, then, Popper and Duhem agree about the influence of metaphysics on science, but they differ in an interesting fashion about one particular example – that of atomism. For Popper, atomism is one of the most striking examples of how a metaphysical theory can have a beneficial influence on science. Duhem, however, held that the influence of atomism on science was a baleful one. Indeed, he frequently attacks atomism, as, for example, in the following passage:

Consider someone, for instance, who would take physical theory just as we have it, in the year of grace 1905, presented by the majority of those who teach it. Anyone who would listen closely to the talk in classes and to the gossip of the laboratories without looking back or caring for what used to be taught, would hear physicists constantly employing in their theories molecules, atoms, and electrons, counting these small bodies and determining their size, their mass, their charge. By the almost universal assent favouring these theories, by the enthusiasm they raise, and by the discoveries they incite or attribute to them, they would undoubtedly be regarded as prophetic forerunners of the theory destined to triumph in the future. He would judge that they reveal a first draft of the ideal form which physics will resemble more each day; and as the analogy between these theories and the cosmology of the atomists strikes him as obvious, he would obtain an eminently favourable presumption for this cosmology.

How different his judgment will be if he is not content with knowing physics through the gossip of the moment, if he studies deeply all its branches, not only those in vogue but also those that an unjust oblivion has let be neglected, and especially if the study of history by recalling the errors of past centuries puts him on his guard against the unreasoned exaggerations of the present time!

Well, he will see that the attempts at explanation based on atomism have accompanied physical theory for the longest time; . . . he will see them constantly being reborn, but constantly aborted; each time the fortunate daring of an experimenter will have discovered a new set of experimental laws, he will see the atomists, with feverish haste, take possession of this scarcely explored domain and construct a mechanism approximately representing these new findings. Then, as the experimenter's discoveries become more numerous and detailed, he will see the atomist's combinations get complicated, disturbed, overburdened with arbitrary complications without succeeding, however, in rendering a precise account of the new laws or in connecting them solidly to the old laws; . . . It will appear clearly to him that the physics of atomism, condemned to perpetual fresh starts, does not tend by continued progress to the ideal form of physical theory. (1905, pp. 303–4)

Almost everyone would now agree with Popper's view that atomism had a beneficial influence on science and disagree with Duhem's claim that the influence of atomism was negative. None the less, this polemic against atomism is of great importance in understanding Duhem and his intellectual work. Atomism as a metaphysical view of the world led naturally, at that time, to a preference for mechanical explanations in terms of concrete visualizable particles. Duhem strongly opposed this approach to physics, advocating instead the use of abstract mathematical theories. He puts his position with customary clarity as follows: 'A physical theory . . . is a system of mathematical propositions, deduced from a small number of principles, which aim to represent as simply, as completely, and as exactly as possible a set of experimental laws' (1904–5, p. 19).

It is not difficult to see the ideological factors behind this dispute. Materialism, mechanism, and atomism were favoured by Duhem's liberal, republican, and anticlerical opponents, whereas Duhem, as a devout Catholic, inclined towards idealism and mathematics. It was Duhem's misfortune that, during the best years of his scientific life, the metaphysical position which he abhorred – namely, mechanism and atomism – led to the most striking advances in physics. This, indeed, affords a partial explanation of Duhem's lack of scientific good sense (*le bon sens*). His particular metaphysical outlook inclined him to theories and approaches which were inappropriate to the most important scientific problems of his day. No doubt there are many other cases in which an inappropriate metaphysics can explain a lack of scientific good sense.

It would be wrong, however, to conclude that metaphysical orientation gives a full explanation of scientific good sense (or the lack of it). In fact, such an explanation is not adequate even in the case of Duhem. Duhem's admiration for abstract mathematical theories should have led him to support Maxwell's theory of electromagnetism and Einstein's theory of relativity, whereas he rejected both in harsh terms. Admittedly, in the case of Maxwell, he might have been misled by the mechanical models which Maxwell employed in his own account of the theory. However, Duhem should have realized that these models were inessential; indeed, he does quote Hertz's famous statement that 'Maxwell's theory is the system of Maxwell's equations' (1904–5, p. 80). As we saw in 5.4, Duhem's erroneous view that arithmetic and geometry are based on common-sense knowledge was one of the reasons why he rejected non-Euclidean geometry and Einstein's theory of relativity.

Although Duhem might not have appreciated this point himself, the development of the key theories of modern physics (relativity and quantum mechanics) did to some extent bear out his polemic against atomism. Consider, for example, the change from Bohr's theory of the atom to the quantum mechanics of Heisenberg, Schrödinger, and Dirac – a change which took place from 1926 on. In the earlier phase of the Bohr atom, models were constructed which involved minute, but still visualizable, particles moving under the action of electrical and mechanical forces. The Bohr atom was conceived of as a miniature solar system. With the quantum-mechanical revolution, everything dissolved into systems of equations, whose solution gave the correct answer but whose interpretation was far from clear. At all events, it was no longer possible to think of concrete particles of the classical type. (For an interesting recent account of these developments, see chapter 4 of Arthur I. Miller's book *Imagery in Scientific Thought* (1984), entitled 'Redefining Visualizability'.)

Now one section of Duhem's polemic against atomism seems curiously prophetic of these developments. It runs as follows:

> Then, as the experimenter's discoveries become more numerous and detailed, he will see the atomist's combinations get complicated, disturbed, overburdened with arbitrary complications without succeeding, however, in rendering a precise account of the new laws or in connecting them solidly to the old laws; and during this period he will see abstract theory, matured through patient labour, take possession of the new lands the experimenters have explored, organize these conquests, annex them to its old domains, and make a perfectly coordinated empire of their union. (1905, p. 304)

Duhem died in 1916, but, even if he had lived to 1926, it is to be doubted whether he would have welcomed the advent of quantum mechanics. After all, he witnessed a similar change from Lorentz's theory of electrons to Einstein's theory of relativity, and, as we have seen, he did not appreciate Einstein's work. We should not dwell on Duhem's failings as a scientist, however, but rather concentrate on his brilliant philosophical insights, which make him one of the greatest philosophers of science of the twentieth century.

Let us now try to see what general conclusions we can draw from this example. We are dealing here with two approaches to the construction of scientific theories, based on different metaphysical ideas. These approaches could be called: (i) concrete models and (ii) abstract mathematics. Concrete models worked wonderfully well in the development of atomic physics up to about 1925, but proved ineffective thereafter, giving way to abstract mathematics. Abstract mathematics did not work well in this field before 1925, as, among other things, Duhem's own scientific work demonstrated. Conversely, concrete models were ineffective after 1925. The example is in fact similar to that of Cartesian metaphysics which we considered earlier. Cartesianism, it will be remembered, proved to be an obstacle to scientific progress during the period when the Newtonian theory of gravity was emerging, but, at a later stage, was helpful for the development of hydrodynamics and Maxwell's theory of magnetism. Similarly, concrete models have worked well in some scientific contexts, abstract mathematics in others. Indeed, the two approaches continue today in different branches of science. Biochemistry, for example, works almost exclusively with concrete models, using little or no abstract mathematics. The models are chemical rather than mechanical. Thus, for example, Mitchell's chemiosmotic hypothesis (see chapter 2, note 3) explains energy flows across cell boundaries by means of an ingenious combination of chemical reactions and ion transport whose net effect is the energy flow. In particle physics, on the other hand, almost everything (too much, one is almost tempted to say!) consists of abstract mathematics.

The conclusion seems to be inescapable that metaphysical ideas are not only meaningful, but necessary for science. They provide an indispensable framework within which specific scientific theories can be constructed and compared with experience. Metaphysics acts as a guide, or heuristic, for science. But while a metaphysical guide is necessary to move in any direction at all, such a guide can just as easily lead in the wrong, as in the right, direction. The very same metaphysical system (whether Pythagoreanism, mechanical materialism, Cartesianism, or whatever) can promote scientific progress in one context or problem situation while acting as an obstacle to science in another. What this shows is that there is no magic formula for doing good science. It is often necessary in scientific research to explore false trails so that the correct one can be found.

9.4   Duhem's Defence of Religion

Duhem was an ardent Catholic, and we cannot give a satisfactory account of his views on philosophy of science without examining their implications for religion. In 1904, Abel Rey published an article on Duhem's philosophy of science in which he argued that this was the philosophy of a believer. Duhem replied to Rey in his 1905 article entitled 'Physics of a Believer'. Here Duhem denies that his philosophy of physics was developed with an apologetic intent, and insists instead that it was 'forced on the author outside of any metaphysical or theological concern, and almost despite himself, through the daily practice and teaching of the science' (1905, p. 275). There is certainly some truth in this. Duhem devoted nearly all his time to the study and teaching of physics and its history and to research in physics. Many of his ideas in the philosophy of science arose from these activities, and are of great interest apart from any religious questions. Yet, if Duhem is telling the truth, he is not telling the whole truth, for his philosophical ideas were also influenced by religious considerations. Indeed, his 1905 article is a proof of this, for in it he uses his philosophy of science to provide a defence of religion.

Duhem, as we have seen, demarcates science from metaphysics. Science has its own sphere and its own methods. Scientific theories have their own subject-matter and general character. The same can be said of metaphysics, but the sphere and methods of metaphysics and the subject-matter and general character of metaphysical theories are quite different from those of science and of scientific theories. Now in Duhem's day it was often maintained that religion had been disproved, or at least undermined, by the advance of science. Duhem argues that religion belongs to the sphere of metaphysics, which is quite different from and does not interact with that of science. So the idea of science undermining religion (or any metaphysical view for that matter) is simply a mistake. As Duhem himself puts it:

> It has been fashionable for some time to oppose the great theories of physics to the fundamental doctrines on which spiritualistic philosophy and the Catholic faith rest; these doctrines are really expected to be seen crumbling under the ramming blows of scientific systems. . . .
>
> Now the system we have expounded gets rid of the alleged objections that physical theory would raise to spiritualistic metaphysics and Catholic dogma; it makes them disappear as easily as the wind sweeps away bits of straw, for according to this system these objections are, and can never be anything but, misunderstandings. (1905, p. 283)

This sounds like a plausible defence of religion, and it might have been expected that Duhem's arguments would be very welcome to the Catholic Church; but, as Martin has shown in his 1991 book on Duhem (see particularly chapter 3), this was not the case. Far from the Catholic Church accepting Duhem's defence of religion, his arguments were thrown under the suspicion of heresy. As Martin says:

> In the encyclical *Pascendi Dominici Gregis* of 1907, two years after Duhem's 'Physics of a Believer', the official position was made clear in the name of Pope Pius X. Of the dangerous aspects of the heresy it called 'modernism' identified by the encyclical two concern me here,

what it called the 'agnosticism' of the modernists, and the separation
of science and faith. The first for example was dangerous because of
the damage it did to natural theology . . . The second was under
suspicion of fideism. (1991, pp. 38–9)

Natural theology is the attempt to argue for theological doctrines
from the character of the natural world. But if, as Duhem claimed,
the spheres of science and religion are quite separate, then natural
theology is hardly a viable enterprise. Fideism is the belief that faith
rests on faith and nothing else, and this does indeed seem to be
supported by Duhem's separation of science and religion. Indeed, as
we pointed out in 8.4, the Catholic Church had grounded its
religious teachings in a particular metaphysical system, the Aristotelianism of St Thomas Aquinas. The aim of orthodox Catholic
intellectuals should then have been to try to reconcile Thomism
with modern science rather than to argue that the two belong to
different spheres.

So much, then, for the attitude of the Catholic Church to
Duhem's defence of religion. Let us now examine this defence in the
light of some of our earlier discussions. Duhem's line of thought on
this subject has a good deal in common with that of Wittgenstein in
the *Tractatus* (see 8.4). Wittgenstein there makes a sharp distinction
between meaningful science and meaningless metaphysics. Religious
doctrines are metaphysical, and so cannot be expressed in meaningful
language, though someone can become aware of their truth through
a mystical experience. Duhem shares with Wittgenstein the idea of
demarcating science sharply from metaphysics, though Duhem, of
course, regards metaphysics as meaningful. In both cases, assigning
science and religion to separate, non-interacting spheres means that
advances in science cannot undermine religious belief.

The difficulty with Wittgenstein's defence of religion in the
Tractatus is that a large number of metaphysical theories do seem to
be perfectly meaningful. The difficulty with Duhem's defence is that
the spheres of science and metaphysics do seem to interact strongly,
and cannot therefore be regarded as entirely separate. As we have
argued earlier in this chapter, metaphysical ideas do influence the
development of science, while advances in science do make some
metaphysical views more plausible than others. Given this situation,
changes in science can have an effect on the credibility of religious
beliefs, contrary to what Duhem claims. As a matter of fact,
Duhem's own writings are not really consistent on this point. In his
defence of religion, he maintains that science and metaphysics are

separate, non-interacting spheres. Yet, in his detailed analyses of historical developments in science, he gives many striking instances of the influence of metaphysical ideas on science, as we saw in the third section of this chapter. Indeed, in his 1905 article 'Physics of a Believer', he actually suggests a way in which some recent developments in science might support a particular metaphysical view of the world.

Duhem says that we can argue in favour of what he calls a 'cosmology' – that is to say, a metaphysical view of the world – by analogy with theories in physics. However, we must be careful to use, not the present state of physical theory, but rather the ideal form towards which it tends. The present state of physics in 1905 was characterized by the triumph of atomism, against which Duhem polemicized. The ideal form towards which physics was tending, according to Duhem, was general thermodynamics. Duhem goes on to develop an interesting and quite plausible analogy between general thermodynamics and Aristotelianism, and concludes:

> If we rid the physics of Aristotle and of Scholasticism of the outworn and demoded scientific clothing covering it, and if we bring out in its vigorous and harmonious nakedness the living flesh of this cosmology, we would be struck by its resemblance to our modern physical theory; we recognize in these two doctrines two pictures of the same ontological order, distinct because they are each taken from a different point of view, but in no way discordant. (1905, p. 310)

It seems here as if Duhem has almost forgotten his 'two separate spheres' justification, given earlier in the same article, and has instead adopted the more orthodox Catholic position of trying to reconcile modern science with the scholasticism on which Catholic doctrine is based. Yet the very way in which he develops this orthodox approach indicates its dangers. If general thermodynamics had indeed triumphed in physics, then this might have been used as an argument for scholasticism. However, by the same token, the failure of general thermodynamics and the success of atomism should be considered as undermining scholasticism, and hence Catholic doctrine.

10

Falsificationism in the Light of the Duhem−Quine Thesis

10.1 Falsificationism and the Falsifiability Criterion

Falsificationism is a theory of scientific method − more specifically, the theory that science proceeds through conjectures and refutations. Now for this theory to work, it is clear that the conjectures put forward by scientists must be capable of being refuted by observation and experiment. This suggests that we should allow as scientific only those conjectures which are refutable or falsifiable. Indeed, Popper himself says: 'But I shall certainly admit a system as empirical or scientific only if it is capable of being *tested* by experience. . . . I shall require that its logical form shall be such that it can be singled out, by means of empirical tests, in a negative sense: *it must be possible for an empirical scientific system to be refuted by experience*' (1934, pp. 40−1). We can see, then, a connection between falsificationism *as a methodology* and falsifiability *as a demarcation criterion* designed to distinguish science from metaphysics.

Yet this connection is a problematic one, since any criticism of falsifiability as a demarcation criterion (and there have been many) would seem to cast doubt, at the same time, on falsificationism as a methodology. My aim in this chapter is to investigate the range of problems which arise out of this situation. I will begin by considering three standard objections to falsifiability as a demarcation criterion. I will argue that the first two of these difficulties (which concern pure existential statements and probability statements) can be surmounted and that the net result of the discussion is, in fact, to provide more arguments in favour of falsifiability. The situation is different regarding the third objection based on the Duhem−Quine thesis. I will argue that this difficulty shows falsifiability to be, if not entirely wrong, at least inadequate.

Quine's reaction to this situation was to deny that any adequate demarcation between science and metaphysics can be drawn at all. Yet Duhem himself both formulated his thesis *and* continued to assume throughout his book that a distinction could be made between science and metaphysics. In the fifth section of this chapter, I will attempt to defend Duhem's rather than Quine's position, by suggesting a new demarcation criterion based on *confirmability*, rather than *falsifiability*. To illustrate this criterion, I will show that it leads to the intuitively plausible result that Newtonian mechanics is scientific, whereas Adler's theory of the inferiority complex is metaphysical. Then in the sixth section I will consider how much of falsificationism as a methodology can be retained if the demarcation criterion is changed in the manner suggested. Rather pleasingly, it turns out that much of falsificationism retains its validity, so that, even assuming our new demarcation criterion, it can still be asserted that scientists should use their creative powers in formulating bold conjectures about the natural world, and that these conjectures should be put in falsifiable form and tested as severely as possible. This still differs from Popper's original conception, in that the refutation of a falsifiable form of a conjecture does not necessarily imply the refutation of the original conjecture itself, so that a high-level theory may be eliminated by a series of defeats rather than by a single knock-out blow.

10.2 Existential Statements

As we saw in 8.5, existential statements such as 'There is (or there exists) a white raven' are verifiable, but not falsifiable. We can verify the claim that there is a white raven simply by observing a raven of that colour; but no finite set of observations of ravens can possibly falsify this claim. Now this logical point might be used as an objection to falsifiability as a demarcation criterion. After all, there do seem to be *bona fide* scientific statements which are existential in character: for example, that there exist egg-laying mammals, or, an example which Popper himself gives (1934, p. 69), that there exists an element with the atomic number 72. It could be argued, therefore, that falsifiability is too strong, and that it inadvertently eliminates some genuinely scientific statements.

Popper's answer to this difficulty is that existential statements on their own are indeed metaphysical. However, with some qualifications, which are usually implicit in the scientific context in which they occur, they become falsifiable. Thus 'There exist egg-laying mammals in this particular (quite closely circumscribed) region of Australia' becomes falsifiable, since we could search the specified region with care and find no such mammals. In terms of his own example, this is how Popper puts the point:

'There exists an element with the atomic number 72' . . . is scientific as part of a highly testable theory, and a theory *which gives indications of how to find this element*. If, on the other hand, we took this existential statement in isolation, or as part of a theory which does not give us any hint as to how and where this element can be found, then we would have to describe it as metaphysical simply because it would not be testable. (1983, pp. 178–9)

This seems to me to be reasonably convincing, and I conclude that existential statements do not constitute an insuperable difficulty for the falsifiability criterion.

10.3   Probability Statements

There is a difficulty connected with the falsifiability of probability statements which Popper himself states very clearly as follows:

The relations between probability and experience are also still in need of clarification. In investigating this problem we shall discover what will at first seem an almost insuperable objection to my methodological views. For although probability statements play such a vitally important role in empirical science, they turn out to be in principle *impervious to strict falsification*. Yet this very stumbling block will become a touchstone upon which to test my theory, in order to find out what it is worth. (1934, p. 146)

To see why probability statements cannot be falsified, let us take the simplest example. Suppose we are tossing a bent coin, and postulate that the tosses are independent and that the probability of heads is p. Let prob.(m/n) be the probability of getting m heads in n tosses. Then we have

$$\text{prob}(m/n) = {}^{n}C_{m}p^{m}(1 - p)^{n-m}$$

The exact meaning of this formula is not important. The important point is that however long we toss the coin (that is, however big n is) and whatever number of heads we observe (that is, whatever the value of m), our result will always have a finite, non-zero probability. It will not be strictly ruled out by our assumptions. In other words, these assumptions are 'in principle *impervious to strict falsification*'.

Popper's answer to this difficulty consists in an appeal to the notion of methodological falsifiability. Although, strictly speaking, probability statements are not falsifiable, they can none the less be used as falsifiable statements, and in fact they are so used by scientists. He puts the matter thus: 'A physicist is usually quite well able to decide whether he may for the time being accept some particular probability hypothesis as "empirically confirmed", or whether he ought to reject it as "practically falsified"' (1934, p. 191).

I have worked out a particular version of this approach in some detail in my 1971 article 'A Falsifying Rule for Probability Statements' and subsequent book *An Objective Theory of Probability* (1973) (cf. part III, pp. 161–226). The full solution involves a great deal of

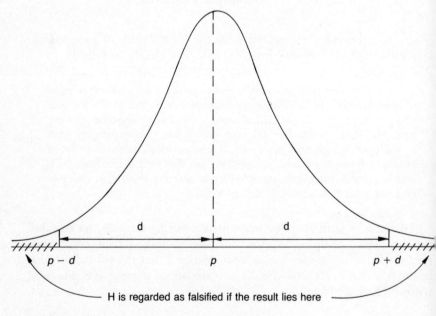

Figure 10.1   Falsifying a probability hypothesis

technical mathematical probability and statistics – particularly a consideration of the theory of statistical testing. However, the basic idea is not difficult, and can be explained as follows in terms of the coin-tossing example.

Although, as already pointed out, each value of m/n has a finite, non-zero probability, some of these probabilities are much higher than others. For example, if m/n is very near to p, prob(m/n) is much higher than it is if m/n is very far from p. Collecting these high probabilities together, we obtain an interval round p, $(p - d, p + d)$ say, such that there is a probability of more than, say, 95 per cent that m/n is within the interval $(p - d, p + d)$, and a probability of less than 5 per cent that m/n is outside this interval. We then regard the underlying hypothesis, H, as confirmed if the observed result m/n lies in the interval $(p - d, p + d)$ and falsified if the observed result lies outside this interval. This procedure can be described as 'cutting off the tails', from the graphical illustration given in figure 10.1.

An actual example is given in my 1973 book (pp. 124–7). I tossed an ordinary coin (an old penny) 2,000 times. Calculations show that, if we assume prob(heads) $= 1/2$, there is a probability of 97.3 per cent that m/n will lie in the interval $(0.475, 0.525)$. The observed value of m/n was actually 0.487, giving a confirmation, rather than falsification.

When developed in full mathematical detail, the procedure agrees reasonably well with the standard theory of statistical testing developed by the Pearsons (father and son), 'Student' (W. S. Gosset), Fisher, and Neyman. There are, however, some rather technical difficulties connected with the so-called one-tailed tests and (particularly) the Neyman paradox. But solutions can be found for these difficulties, and so falsificationism can provide a simple and reasonably satisfactory account of probability which accords well with mainstream statistical practice.

It should be added, however, that there is no consensus regarding the foundations of statistics, and some schools of thought – notably the Bayesian – adopt an approach which is very different from falsificationism. In an important recent book, Howson and Urbach (1989) criticize falsificationism in statistics and defend the Bayesian approach, while in my review of their book (Gillies, 1990), I naturally defend falsificationism against their attacks. Perhaps the most judicious conclusion is that every school of statistics has its problems, and that those of the falsificationist approach are no worse than what is the norm in this field. At all events, probability

statements cannot be regarded as a fatal stumbling block for
falsifiability.

10.4   Falsifiability and the Duhem–Quine Thesis

Falsifiability has been able to overcome the difficulties so far considered. Much more serious, however, are a number of problems
connected with the Duhem–Quine thesis, which was described in
chapter 5. These difficulties do show, as I will now argue, that
falsifiability is not adequate as a demarcation criterion. My intention
is not, however, entirely negative. I will try in the next section
to formulate another demarcation criterion, based on *confirmability*,
rather than *falsifiability*, and will then argue, in the following
section, that, within the new framework proposed, many of the
basic ideas of falsificationism can be retained. In particular, it will
turn out that the falsificationist approach to statistics mentioned in
the preceding section is still viable. Let me begin, however, with
my critique of falsifiability. This is perhaps best approached by
considering again Popper's basic criticism of verifiability, which
he set out clearly in the following passage: 'My criticism of the
verifiability criterion has always been this: against the intention of
its defenders, *it did not exclude obvious metaphysical statements; but it did
exclude the most important and interesting of all scientific statements*, that is
to say, the scientific theories, *the universal laws* of nature' (1963, p.
281).

Let us see if we can turn this criticism against Popper himself.
First of all, does falsifiability fail to exclude obvious metaphysical
statements? Unfortunately the answer is 'Yes', because of a result
known as the 'tacking paradox'. Let T be a falsifiable theory – for
example, Kepler's first law. Let M be an arbitrary metaphysical
statement – for example, 'The Absolute is sleepy'. Set $T' = T \& M$.
Then T' is, like T, falsifiable. Since T is falsifiable, there is an
observation statement O such that O follows logically from T. So if
we find by observation that O is false, it follows that T is false. But
since O follows logically from T, it also follows logically from
$T \& M$. So $T' = T \& M$ is falsifiable. In other words, given a
falsifiable theory, we can always 'tack on' an arbitrary metaphysical
statement, and still have a falsifiable theory. What can be done about
this situation?

My own view is that falsifiability must be supplemented here by considerations of *simplicity*. This might be done in the following way. Let us say that a theory T is of *adequate simplicity* if no theory U can be found such that

(a) U is as simple or simpler than T, and
(b) all the observable results derivable from T are likewise derivable from U.

We then say that a theory, T, is scientific if it is falsifiable and of adequate simplicity. This definition suffices to eliminate cases like T & M which are clearly not of adequate simplicity. However, the problem of eliminating metaphysical components from scientific theories (as it might be called) is no mere philosopher's fantasy, but can crop up in scientific situations where it is not altogether easy to resolve. For example, Newton introduced the concepts of absolute space and time. Were these concepts genuinely scientific, or were they, as Mach claimed, merely metaphysical additions to the theory? The answer is by no means clear.

Let us turn to the second part of Popper's criticism of verifiability: namely, that '*it did exclude the most important and interesting of all scientific statements*, that is to say, the scientific theories, *the universal laws* of nature' (1963, p. 281). Now falsifiability does allow into science some universal laws, including many that are concerned with observable entities and properties. Unfortunately, however, it appears to exclude higher-level theoretical laws, which are normally counted as part of science. To see this, let us take an example considered earlier: namely, Newton's first law, which states that all bodies continue in a state of rest or uniform motion in a straight line unless disturbed therefrom by a force. The difficulty here, as we saw in connection with our discussion of Poincaré's conventionalism in chapter 4, is the following. If we observe a body which is neither at rest nor in uniform motion in a straight line and is apparently not acted on by any force, then we can always postulate an invisible force acting on the body. This is exactly what Newton did in the case of the planets, which move in ellipses, rather than straight lines. He postulated that these planets were acted on by the invisible force of universal gravitation. (see Figure 3.1) A similar device could be applied to any apparent exception to Newton's first law (the law of inertia), which thus does not appear to be scientific according to Popper's falsifiability criterion.

The problem here is, as we saw in chapter 5, posed by the Duhem–Quine thesis. Newton's first law cannot be tested on its own, as an isolated hypothesis, but only as part of a theoretical group. Let us call Newton's first law T. To obtain observable consequences from Newton's theory, we have to add to T:

(1) further theoretical assumptions: namely, the second and third laws of motion and the law of gravity – call the conjunction of these T', and

(2) auxiliary assumptions: for example, that non-gravitational forces have no appreciable effect on the motion of the Sun and the planets, that the mass of the Sun is very much greater than the mass of any planet, and so on – call the conjunction of these A.

Now, from the conjunction T & T' & A, we can indeed deduce observable consequences regarding the motions of the planets. Call one such consequence O. Suppose now that we observe not-O. We cannot conclude 'not-T', but only 'either not-T or not-T' or not-A'. To put the point another way, we know that one of the assumptions used in the deduction is false, but we do not know which one. So none of the assumptions has been falsified. Moreover, as the Duhem–Quine thesis holds for any application of Newton's first law to explain observable phenomena, it follows that this law is unfalsifiable.

How does Popper himself deal with this difficulty? It will be remembered that in formulating his view of falsifiability as a demarcation criterion, Popper speaks not of scientific theories, but of 'theoretical systems'. Interestingly, the term 'system' (in French *système*) is also used by Duhem in this context, as the following passage, already quoted in chapter 5, shows: 'For it is not between two hypotheses, the emission and wave hypotheses, that Foucault's experiment judges trenchantly; it decides rather between two sets of theories each of which has to be taken as a whole, i.e. between two entire systems, Newton's optics and Huygens' optics' (1904–5, p. 189). An appeal to this notion of 'theoretical systems' does indeed constitute Popper's answer to the problem posed by the Duhem–Quine thesis, as the following passage shows:

> More serious is an objection closely connected with the problem of *context*, and the fact that my criterion of demarcation applies to *systems of theories* rather than to statements out of context. This objection may be put as follows. No single hypothesis, it may be said, is falsifiable, because every refutation of a conclusion may hit any single

premise of the set of all premises used in deriving the refuted conclusion. The attribution of the falsity to some particular hypothesis
that belongs to this set of premises is therefore risky, especially if we
consider the great number of assumptions which enter into every
experiment. . . . The answer is that we can indeed falsify only *systems
of theories* and that any attribution of falsity to any particular statement within such a system is always highly uncertain. (1983, p. 187)

Let us see how this works out in terms of our example of
Newton's first law, T. As we have seen, T is not falsifiable, but the
conjunction of T & T′ & A is falsifiable. One option for Popper
would therefore be to say that T is not scientific but metaphysical,
whereas T & T′ & A is scientific. This suggestion is not very
satisfactory from Popper's own point of view, since he has always
strongly criticized the verifiability criterion on the grounds that it
excluded the universal laws of nature. Now given the view we are
considering, Newton's first law, a classic paradigm of an important
and interesting law of nature, is excluded as a piece of metaphysics.
The present suggestion could not, therefore, be considered as satisfactory by Popper or, indeed, by myself.

There is, however, another possibility. Could we not say that a
theory T_1 is scientific if it is part of one or more systems of theories
of the form T_1 & T_2 & . . . & T_n which are falsifiable? This
modified demarcation criterion certainly allows Newton's first law
to be scientific; but, unfortunately, it has the consequence that any
arbitrary metaphysical statement is scientific. This can be shown by
using what I will call the *Ayer example*. This example was originally
put forward by Ayer in the introduction to his 1946 book (pp.
11–12) as a criticism of his own previous version of the verifiability
criterion; and he then goes on to formulate a new version of
the principle of verification. Ayer notes that he was influenced
by Berlin's criticisms (cf. Berlin, 1939), though Berlin's counter-
example is in fact different from Ayer's. But although the Ayer
example was first proposed in connection with the verifiability
criterion, it can usefully be considered in connection with the
falsifiability criterion, and also in connection with the confirmability
criterion to be introduced in the next section.

Let M be an arbitrary metaphysical statement – for example, 'The
Absolute is sleepy'. Let O_1 . . . O_n be any arbitrary observation
statements – for example, 'This is white', 'This is a pen', and so
forth. Then M is a component of the following falsifiable systems:
M & (if M, then O_1), M & (if M, then O_2), . . . M & (if M, then O_n).

Hence, by our modified criterion, M ought to be scientific, which it is not. I conclude that falsifiability is not adequate as a demarcation criterion.

10.5 A Suggested Demarcation Criterion Involving the Principle of Explanatory Surplus

If falsifiability is inadequate as a demarcation criterion, the next obvious possibility to consider is *confirmability* or *corroborability*. The suggestion is that a theory is scientific if and only if it is *confirmable*: that is, capable of acquiring some positive degree of support from a possible observation statement. If a theory (T, say) is falsifiable, then it is certainly confirmable. For let O be an observation statement which is a potential falsifier of T, then, if not-O is observed, this will support T. On the other hand, there are theories or laws such as Newton's first law which are confirmable without being falsifiable. I am thus suggesting the introduction of a new level (level 2) between falsifiable statements (level 1) and metaphysical statements (level 3). If observation statements are classified as level 0, the scheme can be presented as in table 10.1, where, for conciseness, I have omitted considerations relating to simplicity. The arrows connecting levels 1 and 2 to level 0 are designed to indicate that scientific theories of any level, when once confirmed and accepted as background knowledge, can be used in the interpretation of sensory experience which is needed to produce observation statements.

Let us now examine whether it is reasonable to regard Newton's first law, T, as being confirmable, though not falsifiable. T is part of various conjunctions such as T & T' & A which are confirmable; so, if we suppose that the support acquired by the conjunction is distributed among the components of the conjunction, then T will be confirmable. But now it could be objected that this approach will run into the difficulties created by the Ayer example.

Once again, let M be an arbitrary metaphysical statement – for example, 'The Absolute is lazy' – and let O be an arbitrary observation statement – for example 'This is white'. Then we have that O follows logically from M & (if M, then O). It would seem, therefore, that O supports M & (if M, then O); hence, again supposing that the support of the conjunction is distributed among the components of the conjunction, M is supported to some extent

Table 10.1   Classification of statements into levels 0, 1, 2, and 3

Level	Status	Criterion	Example
3	Metaphysical	Not confirmable	Greek atomism
2	Scientific	Confirmable, but not falsifiable	Newton's first law
1	Scientific	Falsifiable and confirmable	Kepler's first law
0	Observation statement	Truth-value determinable by observation	Statement recording position of Mars at a particular time

by O. Thus any arbitrary metaphysical statement would appear to be confirmable, and our whole attempt to demarcate science from metaphysics on the grounds of confirmability appears to collapse.

To avoid this difficulty, I will suggest a principle which limits the circumstances in which evidence can be considered as supporting a theory, and, which, in particular, blocks the confirmation of M in the Ayer example. To illustrate what is involved here, let me first give an example of a general principle of confirmation which would, I think, be accepted by most philosophers of science. This is the 'principle of severe testing' which states that the more severe the tests which a hypothesis h has passed, the greater is the confirmation of h. Popper seems to have been the first to formulate this principle, and he puts it as follows: 'It is not so much the number of corroborating instances which determines the degree of corroboration as *the severity of the various tests* to which the hypothesis in question can be, and has been, subjected' (1934, p. 267).

My aim, then, is to introduce a principle of confirmation theory having the same logical status as, though a different content from, Popper's principle of severe testing. This principle will be called *the principle of explanatory surplus*. It is intended to supplement, rather than replace, Popper's principle of severe testing, with which it is entirely compatible. The idea behind it is this. The principle denies that if e follows logically from h, this automatically means that e supports h. Not all facts which follow from a given hypothesis support that hypothesis, so the principle claims, but only a subset of these deducible facts – a subset which constitutes an explanatory surplus. In particular, it will transpire that O does not support M & (if M, then O). I will now attempt to formulate the version of the principle of explanatory surplus which applies to the kind of case we have been considering.[1]

Let us suppose that a scientist is trying to give a theoretical explanation of a set of facts which I will denote by $f_1 \ldots f_n$. To do so, he or she makes a number of theoretical assumptions which I will denote by $T_1 \ldots T_s$. The facts in question may be concerned with single events or may be universal laws connecting observables, such as 'All ravens are black' or Kepler's laws. In either case we assume that $f_1 \ldots f_n$ are well confirmed by observation and experiment, and so can be assumed to be true (at least when interpreted as approximations) while the attempt at theoretical explanation is being made.

Let us now further suppose that each of $f_1 \ldots f_n$ follows logically from some subset of $T_1 \ldots T_s$ together with initial conditions which are established by observation and experiment. So our assumption is that for each i $(1 \leq i \leq n)$, f_i follows logically from a *theoretical system* of the form O_i & T_{i1} & $\ldots$ & T_{ip}, where O_i is an observation statement, and each T_{ij} $(1 \leq j \leq p)$ is a member of the set $T_1 \ldots T_s$.

The question before us is: Given this general situation, to what extent, if any, are $T_1 \ldots T_s$ supported by $f_1 \ldots f_n$? The principle of explanatory surplus states that $T_1 \ldots T_s$ are supported not by all the facts they explain, but only by that fraction of the facts which can be considered an explanatory surplus. There is implicit here an economic analogy. The successful theoretician is like a successful entrepreneur. To be successful, an entrepreneur has to choose investments $I_1 \ldots I_s$ for his or her capital in such a way that he or she obtains a surplus, and the bigger this surplus, the more successful the entrepreneur. Likewise, to be successful, the theoretician has to choose the theoretical assumptions $T_1 \ldots T_s$ in such a way as to generate a surplus, and the bigger this surplus, the more successful the theoretician. In the first case, the surplus is an economic one, and takes the form of an excess of receipts over outlays. In the second case, we are concerned with an explanatory surplus, which consists, roughly speaking, of an excess of facts explained over theoretical assumptions employed.

Next we have to consider how the explanatory surplus should be estimated. The simplest and most straightforward method is to subtract the number of theoretical assumptions used from the number of facts explained. So, in our notation, the size of the explanatory surplus is estimated as $n - s$. Thus, if a theoretician has to explain n facts, and needs to postulate n or more than n theoretical assumptions to do so, then the facts do not support the theoretical assumptions at all, even if the theoretical assumptions do explain the facts (in the sense of the deductive model of explanation). A theoretician in such a situation is like an unsuccessful entrepreneur, who either breaks even or makes a loss, but in any case fails to make a profit.

Let us see how the principle of explanatory surplus deals with the Ayer example. Here we have that O follows logically from M & (if M, then O), and the question is whether O supports M & (if M, then O). Well, we have two theoretical assumptions M and (if M, then O), but only one fact is explained – namely, O. So there is an explanatory deficit rather than an explanatory surplus, and therefore, according to the principle of explanatory surplus, O does not support M & (if M, then O).

I will now show that we can deal with the tacking paradox using the principle of explanatory surplus. Let T be a *bona fide* scientific theory, and let us tack on to T an arbitrary metaphysical assumption M to obtain T & M. Intuitively we prefer T to T & M, but why? If we accept the principle of explanatory surplus, then T is always better supported than T & M by given evidence, for, since T & M has one more theoretical assumption (namely, M) than T, the explanatory surplus generated by T & M will always be smaller than that generated by T. This certainly gives a reason for preferring T to T & M, and, in general, the principle of explanatory surplus motivates the search for theories which are as simple as possible, in the sense of containing as few theoretical assumptions as possible.

Let us next consider an objection which might be made to our method of estimating the size of the explanatory surplus. It could be said that the division into s separate theoretical assumptions $T_1 \ldots T_s$ or n separate facts $f_1 \ldots f_n$ is rather arbitrary. We might, for example, consider the conjunction f_{n-1} & f_n as a single fact f'_{n-1}, thereby reducing the number of facts, and hence the explanatory surplus, by one. It might in some cases be possible to represent f_{n-1} & f_n quite naturally as a single fact. For example, let

$f_{n-1} = $ X is a sibling,
$f_n \quad = $ X is male;

then

$f'_{n-1} = $ X is a brother.

There is undoubtedly a real difficulty here, but it does not, in my opinion, make the suggested method of estimating the explanatory surplus valueless. In a concrete scientific situation, where standard linguistic formulations are in use, there will generally, so I would claim, be a natural way of effecting the division into separate facts or separate theoretical assumptions. Of course, this division will never be completely determinate, but then, in the present context, we are aiming only at a rough qualitative estimate – not at anything precise and quantitative. If we were writing an artificial intelligence program, for example, we might want to make our estimate of the explanatory surplus precise and quantitative, but in that context, the logical language in which everything was formalised would provide a method for distinguishing separate facts and separate theoretical assumptions.

There is another consideration which goes some way towards resolving the problem in hand. In practice, we often want to estimate degrees of confirmation in order to evaluate two competing theories, such as, for example, the Copernican and the Ptolemaic theories at some time in the period 1543–1687. Now in such a case, the precise details of how we make the division into separate facts and separate theoretical assumptions does not matter too much, *provided* it is done in the same way for each of the two competing theories.

This is all I want to say about the problem in general terms. Here, as so often, the proof of the pudding is in the eating. The principle of explanatory surplus has been designed to enable confirmation theory to be applied to actual scientific examples, whether historical or present-day. If in practice it proves possible to estimate the size of the explanatory surplus in a sensible and natural way, and if the principle of explanatory surplus leads to satisfactory results, then there is a strong case for adopting it. If not, not. The evaluation of the principle is thus a matter for further investigation, and I will content myself here with showing very briefly how the principle can be applied in two different cases, one involving a successful theory and the other an unsuccessful theory. These cases are Newton's laws and Adler's theory of the inferiority complex. In the next section we will give as a third example: the early quantum theory of Planck and Einstein.

10.5.1 Newton's Laws

Newton presented his theory as consisting of three laws of motion and the law of gravity. This was a perfectly natural formulation at the time, and we may say, therefore, that his theory divided naturally into four theoretical assumptions: T_1, T_2, T_3, and T_4, where T_1, T_2, and T_3 are the three laws of motion and T_4 is the law of gravity. Perhaps, as suggested earlier, we should add an A, containing the auxiliary assumptions, but, for simplicity, I will omit this. As our estimation of the size of the explanatory surplus is, in any case, rough and qualitative, this omission will not affect the general line of argument.

Let us turn to the facts which Newton sought to explain. Here again, there is a natural division into Kepler's three laws of planetary motion and Galileo's law of falling bodies. Applying the principle of explanatory surplus, we conclude that if Newton had explained Kepler's laws and Galileo's law *and nothing else*, he would not have generated an explanatory surplus, and his theory would not have been supported.

At first sight, this result may seem surprising, but further reflection shows it to be reasonable. Newton produced a complicated theoretical system involving new concepts (force and mass) and a bold and curious assumption concerning gravitational attraction. What would have been the point of adopting such a complicated system if it explained no more than the observational laws of Kepler and Galileo. In such a situation it would surely be better to stick to the observational laws and reject the theory as a piece of metaphysics. This is exactly what the principle of explanatory surplus suggests by assigning zero support to the theory in this case.

Here, however, we are talking merely hypothetically and not in accordance with historical reality. Newton's theory did not explain just Kepler's laws and Galileo's law, but a great deal more besides. In the *Principia* Newton explained, with reasonable success, the laws of impact, the tides, the inequalities of the Moon's motions, and some planetary perturbations. He was also able to derive results concerning the figure of the Earth and comets. We clearly have here a large explanatory surplus, and Newton's theory is correspondingly strongly supported.

This application of the principle of explanatory surplus justifies to a considerable extent some remarks of Duhem's which we quoted earlier in 3.3. This is what Duhem says:

> Therefore, if the certainty of Newton's theory does not emanate from the certainty of Kepler's Laws, how will this theory prove its validity? It will calculate, with all the high degree of approximation that the constantly perfected methods of algebra involve, the perturbations which at each instant remove every heavenly body from

the orbit assigned to it by Kepler's Laws; then it will compare the calculated perturbations with the perturbations observed by means of the most precise instruments and the most scrupulous methods. (1904–5, pp. 193–4)

There are two things in this passage with which I disagree. First of all, it seems to me too strong to speak of a theory *proving its validity*; it would be better to refer to a theory *acquiring a high degree of confirmation*. Evidence can never prove that a scientific theory is true, but it can show that the theory is well confirmed. Secondly, Duhem mentions only the perturbations of orbits; but other things, such as the theory of the tides, the calculation of the paths of comets, and so on, should be included as well. With these reservations, I would support what Duhem says here, which is indeed an instance of the principle of explanatory surplus.

10.5.2 Adler's Theory of the Inferiority Complex

Popper, who worked with Adler for a while, discusses the theory of the inferiority complex briefly in *Conjectures and Refutations* (1963, p. 35). I will here present a variant of one of Popper's examples.

In this example we have two facts: f_1 and f_2. They can be described as follows:

f_1: Mr A is walking beside a river. He sees a child fall in. Without hesitating, he jumps in, and gallantly rescues the child.

f_2: At the same place, but on another occasion, Mr B is walking beside the river. He sees a child fall in. Although he can swim as well as Mr A, he fears that he might drown if he tries to rescue the child, and so walks quietly away. (In order not to make the story too tragic, let us suppose that the child is washed ashore without drowning.)

How are the facts f_1 and f_2 explained on Adler's theory? According to that theory, everyone has an inferiority complex. Some people, however, struggle to overcome this complex by performing difficult and dangerous feats whenever possible. Mr A falls into this class, and his behaviour is thus explained. Other people, however, are totally mastered by their inferiority complex, and will never undertake anything which appears difficult or dangerous because they feel too incapable and inferior to be able to carry out such a task with success. Mr B is such a person, and this explains why he acted as he did. Here, then, are two characteristic Adlerian explanations of observed human behaviour. The question before us is whether such explanations give any support to Adler's theory.

In order to give a reply, we have to analyse the number of theoretical assumptions used in these Adlerian explanations. Clearly we have the assumption, T_1, that all human beings have an inferiority complex. However, to obtain the explanations, this general assumption must be supplemented by two particular assumptions concerning Mr A and Mr B respectively. These are:

T_2:   Mr A struggles to overcome his inferiority complex by performing difficult and dangerous feats whenever possible.

T_3:   Mr B is so mastered by his inferiority complex that he will avoid even attempting anything that appears difficult or dangerous.

In this case, then, two facts (f_1, f_2) are explained by three theoretical assumptions (T_1, T_2, T_3). Thus no explanatory surplus is generated, but rather, an explanatory deficit, and the explanations give no support to Adler's theory of the inferiority complex. Assuming, which I think to be the case, that all Adler's explanations in terms of the inferiority complex are of the above form, it follows that Adler's theory of the inferiority complex is not confirmable, and is hence, by our demarcation criterion, metaphysical.

10.6   How much of Falsificationism can be Retained?

Since we are changing the demarcation criterion from *falsifiability* to *confirmability*, it might be thought that we would have to reject many of the methodological precepts of *falsificationism*; but this turns out not to be the case. In fact, nearly all the prescriptions of falsificationism can be retained; they need to be supplemented, rather than abandoned. This is a very desirable situation, since falsificationism has proved its worth in many branches of scientific enquiry. Let us now see how this comes about.

The main difference between Popper's approach and the one advocated here is the following. Popper operates with a three-level model. He has observation (or basic) statements, scientific laws or theories which are assumed to be falsifiable, and metaphysical statements. Here, however, I have suggested a four-level model. Popper's three levels are identified with my levels 0, 1 and 3, but I include a level 2 (scientific but not falsifiable), which is not in Popper. The key difference, then, is the addition of one extra level.

It follows from this that, as far as level-1 hypotheses or theories are concerned (and there are many such hypotheses and theories in science), Popper's methodology of conjectures and refutations can be maintained unchanged. At this level, hypotheses are put forward as conjectures. They must then be tested as severely as possible. If the tests result in refutation, the hypotheses must be modified. If the tests result in verification, the hypotheses can be accepted provisionally as corroborated, but the need for further testing remains. All these features of falsificationism as a methodology can be retained unchanged for hypotheses at this level. Now if we introduce a *falsifying rule for probability statements* (cf. Gillies, 1971 and 1973), then the majority of statistical hypotheses become falsifiable, level-1 hypotheses. So a falsificationist methodology for statistics can be retained within the present framework. Then again, the present framework contains a level 4 of meaningful metaphysics, so that the claims of Popper and Duhem about general metaphysical ideas acting as a heuristic for the construction of scientific hypotheses can still be supported.

Let us now consider level-2 hypotheses, such as Newton's first law, which are confirmable, so scientific, but not falsifiable. Even as regards theories and hypotheses of this level, many of the precepts of falsificationism can be retained. The aim is to confirm these level-2 hypotheses; but, in order to do so, it is necessary to turn them into level-1 hypotheses which can then be tested as severely as possible. There are a number of ways in which a level-2 hypothesis can be converted into a level-1 hypothesis. It may be possible, by adding some extra assumptions, to derive a level-1 law from the level-2 hypothesis. This level-1 law can then be tested experimentally. We shall give some examples of this in a moment. Then again, the level-2 hypothesis may be converted into a level-1 *theoretical system* by the addition of further theoretical assumptions. The consequences of this theoretical system can then be compared with experience. Note, however, that the confirmation of the level-1 law or the level-1 theoretical system confirms the level-2 hypothesis only if the conditions specified by the principle of explanatory surplus are satisfied. This produces some new and important methodological precepts not to be found in falsificationism. In the transition from level 2 to level 1, scientists are required to use as few theoretical assumptions as possible and to derive as many facts as possible. We shall see in a moment how these precepts were followed by such great scientists as Planck and Einstein.

Since the position here advocated supplements, rather than abandons, falsificationism, it could appropriately be called *modified falsificationism*. I have already suggested (in chapter 5) the use of this term to characterize Duhem's position, but since the present account of science is really nothing other than a systematic development of Duhem's, there is really no contradiction in terminology here. There is one, frequently voiced criticism of falsificationism which is avoided by modified falsificationism. It is often said, particularly by working scientists, that the aim of science is not to prove theories wrong, as falsificationism seems to suggest, but to find theories which actually work. This apparent paradox is neatly resolved by modified falsificationism. The aim of science on the present account is indeed to find theories which are well confirmed, for only such theories can form the basis of satisfactory practical applications. However, in order to get theories which are well confirmed, it is necessary to submit any theory we propose to harsh criticism and severe experimental testing. Only if a theory survives this ordeal can it become well confirmed. Thus criticism and testing are not ends in themselves, but means to an end – that of theories which are well confirmed and can form the basis of practical applications.

Modified falsificationism and its four-level model can also be compared with falsificationism and its three-level model by seeing how well they fare in the analysis of key episodes in the history of science. I think it will be found that modified falsificationism can in most cases furnish a more realistic and convincing account. I will illustrate this by considering just one episode: the introduction of quantum theory by Planck and Einstein in the years 1900–5.[2]

Planck's introduction of quantum theory was occasioned by his study of the problems of heat radiation and, more specifically, of *black body radiation*. A black body is one which absorbs all the radiation falling on it. Such a body is to some extent a theoretical construct, but it can be approximately realized experimentally by a small hole in the wall of an oven at uniform temperature. Any radiation entering such a hole is unlikely to find its way out again, so that the hole acts as a black body. In 1879 Stefan proposed the law that the amount of black body radiation is proportional to the fourth power of the absolute temperature, T. This result was improved by Wien, who in 1893 obtained from general thermodynamical considerations his *displacement law* which gave Stefan's law as a special case. Wien's displacement law relates the quantity of radiation to its wavelength. For a given temperature, there is a definite wavelength (λ_{max}, say) at which the amount of radiation is at its maximum. As the temperature increases, so λ_{max} decreases. This is why we speak of a *displacement* law.

Now Wien's displacement law contains an unknown function which cannot be fixed by general thermodynamical considerations. We can calculate it, however, if we assume a model of the radiating body. The simplest model of a body radiating at a given frequency is a linear harmonic oscillator. If we adopt this model, we can obtain a formula relating the amount of energy radiated at a given frequency to the value of the frequency and the value, T, of the absolute temperature. This is the *Rayleigh–Jeans radiation law*, which agrees very well with experiment for low frequencies (or equivalently long wavelengths), but diverges dramatically for high frequencies (or short wavelengths). In fact, it follows from the Rayleigh–Jeans radiation law that the amount of energy radiated should go to infinity as the wavelength goes to zero. This was called the 'ultra-violet catastrophe'. In fact, however, the amount of energy falls to zero as the wavelength goes to zero. Note that the Rayleigh–Jeans radiation law is a falsifiable law, and, indeed, was falsified in an unmistakable fashion.

As so often in science, the falsification of a law led to an advance, for it was at this point in the story that Planck made his crucial intervention. Planck suggested that the energy of the radiation might not be emitted continuously, but in multiples of some minimal unit of energy, or *quantum*. If this quantum of energy has value e_0, say, the energy emitted by the oscillators can only take values which are multiples of e_0, such as e_0, $2e_0$, $3e_0$, and so on. It is clear that, in formulating this theory, Planck must have been influenced by atomism in a rather abstract sense, so that, contrary to Duhem, we have here another instance of the beneficial influence on science of metaphysical atomism.

On developing his idea, Planck discovered that he had to set

$$e_0 = h\nu,$$

where ν is the frequency of the radiation, and h is a universal constant, now known as *Planck's constant*. From this assumption, it is possible to derive *Planck's radiation law*, which agrees very well with experiment.

Let us review the situation at this juncture. Planck's radiation law is a level-1, or falsifiable, law, and has been tested and confirmed by experiment. This law, however, was derived from a level-2 law postulating that the energy of the oscillators is quantized and that the quantum of energy is hv. This level-2 law has so far only succeeded in explaining a single fact, and so, according to the principle of explanatory surplus, it has not at this stage been confirmed – although undoubtedly it has proved heuristically very fruitful. It was at this point that Einstein took a hand, and very much altered the picture.

The year 1905 was Einstein's *annus mirabilis*. In that year he published three papers (all in the same journal, the *Annalen der Physik*), each of which made a profound contribution to a different branch of physics. One of these papers introduced the special theory of relativity; another dealt with Brownian motion; but it is the third, published in March 1905, which we shall consider here. This paper has the somewhat strange title 'On a Heuristic Viewpoint [*heuristischen Gesichtspunkt*] concerning the Production and Transformation of Light'. We shall see in a moment why Einstein used the phrase 'heuristic viewpoint'.

In this paper Einstein was concerned with a quite different phenomenon from black body radiation. This was the photoelectric effect, which had been discovered by Hertz in 1887 and investigated by Lenard in 1902. Figure 10.2 shows Lenard's experimental

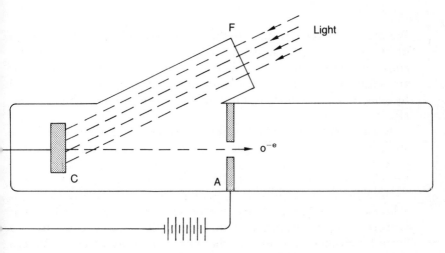

Figure 10.2 Lenard's experimental arrangement for investigating the photoelectric effect. From Max Born, *Atomic Physics* (1935), p. 85.

arrangement. The photoelectric effect occurs when short-wave
(ultraviolet) light falls on certain metals, thereby causing the emission
of electrons. In figure 10.2 light enters a window, F, and strikes
the metal at C. This liberates electrons, which are accelerated (or
retarded) in the field between C and A.

Now one interesting result which emerged was that light below a
certain wavelength, or, equivalently, above a certain frequency, is
needed for any electrons at all to be emitted. The exact value of the
minimum frequency needed for emission depends on the particular
metal. Einstein had the ingenious idea of using Planck's quantum
hypothesis to explain this result. He therefore postulated that the
energy of light – and indeed of any electromagnetic radiation – is
divided into discrete quanta, each containing an energy $h\nu$, where ν
is the frequency of the light, and h is Planck's constant. This is, in
fact, a generalization of Planck's hypothesis, since Planck applied his
quantum idea only to the energy exchange between the oscillators
involved in his model of black body radiation.

Having made this assumption, Einstein further postulated that
when a quantum of light struck an electron in the metal's surface, it
gave up all its energy to the electron. If work of a quantity A needs
to be done for the electron to overcome the forces holding it in the
metal's surface, then the electron will be liberated only if $h\nu > A$,
and it will then move off with kinetic energy E, where

$$E = h\nu - A \tag{1}$$

We can see at once from this that electrons will be emitted only if
the frequency of the light, ν, is greater than A/h, where A is a
constant whose value depends on the metal, and that, if electrons
are emitted, the frequency and kinetic energy will be related
by equation (1), which is Einstein's equation for the photoelectric
effect.

If we consider an apparatus like that of Lenard illustrated in figure
10.2, it is clear that the frequency, ν, of the incident light can easily
be controlled. The velocity, and hence the kinetic energy, E, of
the electrons can be measured by deflection experiments or by a
counter-field. Thus Einstein's equation for the photoelectric effect is
a level-1 law whose validity can be checked experimentally. It is
worth noting, however, that this level-1 law was derived from a
level-2 hypothesis postulating the quantum nature of light. We can
now return to the title of Einstein's paper, and explain the reason
why he spoke of a 'heuristic viewpoint'. The quantum theory of

light was indeed heuristic for Einstein, since it enabled him to derive his level-1 equation for the photoelectric effect. Even such a daring thinker as Einstein was reluctant, however, to regard the level-2 quantum theory of light as anything more than heuristic, for an obvious enough reason. The quantum theory of light contradicted the wave theory of light, which was supported by a wide variety of interference and diffraction experiments and which was a consequence of Maxwell's equations. It must have seemed then that the quantum theory of light was ruled out by theoretical considerations, and could therefore be only a heuristic device.

Einstein himself had doubts about the quantum theory of light, and other scientists had even stronger doubts. Among those who were convinced that the quantum theory of light must be wrong was the famous experimental physicist R. A. Millikan. Millikan reasoned that if the quantum theory of light was wrong, then its consequence, Einstein's equation for the photoelectric effect, was also likely to be wrong. He therefore set out to refute this equation experimentally. After ten years of testing the equation as severely as possible, Millikan was forced to conclude, contrary to his expectations, that the equation predicted exactly the observed results, and to assert its unambiguous experimental verification, in spite of its unreasonableness.

There could hardly be a more striking example of falsificationist principles. Millikan set out to falsify a law which he believed on theoretical grounds to be wrong. He tested the law as severely as he could, and only when his attempts at refutation failed was he forced to conclude that the law had been strongly corroborated. Millikan's decision to attempt to falsify Einstein's equation should not, of course, be taken as showing that Millikan was a bad physicist, but rather that he was a good one. As Cohen astutely remarks: 'Einstein's theoretical suggestion tended to be ignored more than actively combatted. Millikan, a truly great physicist, was an exception to the rule' (1985, p. 426).

Not every scientist who proposes a bold new hypothesis has the good fortune to find a Millikan who is willing to test it severely. It is for this reason that scientists should internalize the prescriptions of falsificationism, and make a point of criticizing and trying to refute their own theories. An example of a scientist who did act in this way is Mitchell. When he first proposed his chemiosmotic hypothesis, this was too far removed from current ideas for any other scientist to think it worth testing, and Mitchell had himself to submit his hypothesis to experimental testing.[3]

Let us now turn our attention from Einstein's level-1 law of the photoelectric effect to the level-2 hypothesis from which it was derived: that is to say, the quantum theory of light, or rather, more generally, of electromagnetic radiation. After Millikan's work, the position was the following. From a single level-2 hypothesis (the quantum theory of electromagnetic radiation), two level-1 laws in different areas of physics had been derived and confirmed experimentally. These were Planck's radiation law and Einstein's law of the photoelectric effect. The identity of the level-2 hypothesis in these two separate cases was demonstrated in a striking fashion by the fact that the same constant, h (Planck's constant), could be determined experimentally in two quite different situations and was found to have the same value. We have, then, a single theory explaining two distinct facts, and so, by the principle of explanatory surplus, the theory becomes confirmed. Here then we have the beginning of the confirmation of the wave–particle duality which was to receive further confirmation in due course. This example shows the great importance of falsificationism and, at the same time, the need to supplement falsificationism by introducing a new level of hypotheses and a principle (the principle of explanatory surplus) showing how such hypotheses can be confirmed.

10.7   Some Concluding Philosophical Remarks

I will conclude this chapter, and indeed the book, with two points of a rather more philosophical nature. The first concerns the question, which arose in connection with Wittgenstein's later theory of language, of the relation between the scientific nature of a theory and its application in practice. Now the attempt to apply a theory in practice can be regarded as analogous to submitting the theory to an experimental test. Bearing in mind the qualifications discussed earlier (the principle of explanatory surplus and so forth), we can say that if a theory passes an experimental test, it is confirmed, while if it fails the test, it is disconfirmed. Analogously, it seems reasonable to say that if a theory is applied successfully in practice, it is confirmed, and so, by our demarcation criterion, must be regarded as scientific. I remarked earlier (8.4) that a theory can be scientific even though it has not been applied in practice. For example, Einstein's general theory of relativity was confirmed by observation, and so shown to be scientific, many years before it was used in any practical application. It seems, however, that the converse holds good; that a theory which is applied successfully in practice is thereby shown to be scientific.

Broadly, then, the successful practical application of a theory can be taken as a sufficient condition for that theory's being scientific. There is, however, an important difficulty in applying this criterion which must now be pointed out. Experiments are performed under very carefully controlled conditions, so that it is usually possible to say whether or not the result was in accordance with the theoretical predictions. By contrast, practical applications often occur in rather disordered and uncontrolled environments, so that the use of a theory may appear to be successful without really being so. This can be illustrated by an incident from the life of the explorer David Livingstone.[4]

Livingstone was a missionary with the Bakwain tribe in Africa in the 1840s. He succeeded in converting the tribe's chief, Sechele, to Christianity. Sechele was, in fact, Livingstone's only convert in his entire career. Now, as Sechele was the chief, he had to perform the rain-making ceremony, an important office in a country with very little rain. After Sechele's conversion, Livingstone persuaded Sechele to give up rain-making as a heathen ritual. Unfortunately, a prolonged drought set in immediately after Sechele's decision, and the tribe eventually forced Sechele to start the rain-making ceremony again. Now, from our modern point of view, we can say that the onset of drought was a coincidence, and had nothing to do with the cancellation of the rain-making ceremonies. However, Sechele's tribe can hardly be blamed for thinking that the events demonstrated the practical efficacy of their rain-making ceremonies, and confirmed the truth of the mythological theories on which these ceremonies were no doubt based.

We should beware, moreover, of assuming that similar difficulties do not exist in our modern society. I have already mentioned (8.6) the difficulty of deciding whether a government's economic policy is really proving successful or not – a situation which politicians readily turn to their advantage. Similar difficulties can occur in medicine. An example is provided by the recent surgical treatments of breast cancer. If a cancer was detected in a breast, it became standard practice to remove the whole breast, on the grounds that any less drastic treatment would leave cancer cells in the patient, which would cause a regeneration of the tumour. Subsequently, however, it was called into question whether such a devastating and disfiguring operation was necessary. It was argued that removing just the tumour itself, though it would indeed leave some cancer cells in the patient, would at the same time leave the patient stronger, both physically and psychologically, and so better able to recover fully from the disease. The exponents of this point of view thus argued that a much less drastic operation would produce a better recovery rate. Of course, in theory it would be possible to test these different views by dividing patients at random into two groups, giving those in one group one operation and those in the other group the other, and then seeing which group had the better recovery rate. However, such random trials on human guinea-pigs raise ethical questions.[5] Is it really moral to give a group of patients a treatment which doctors may believe on general theoretical grounds to be unsatisfactory? The important point is that, without systematic experimentation, which is not always possible, it may be genuinely difficult to tell in some cases whether the application of a theory is proving to be successful or not. Despite these difficulties, the success or failure of practical applications of a theory is certainly a significant factor, and should not be neglected in trying to assess the confirmation of the theory.

So I come to my last point, which is an attempt to relate the science/metaphysics demarcation to the distinction between knowledge and belief. Traditionally (the suggestion goes back at least to Plato's *Theaetetus*) knowledge has been identified with justified true belief. To begin with, I would like to drop the requirement of truth. Almost all the scientific theories introduced before about 1800 have subsequently been shown to be false in some respect or other. Thus, if we insist that knowledge has to be true, it follows that there was no scientific knowledge before about 1800. This I regard as an unacceptable conclusion, so simply require that knowledge be justified belief. If, further, we adopt an *empiricist* position that ultimately there can be no justification except in terms of agreement with observation and experiment, it follows that knowledge can be identified with belief which is well confirmed by experience. This is the definition of knowledge which I would like to adopt. Now, by our demarcation criterion, a metaphysical theory is one which is not confirmable by experience. Hence, *a fortiori*, a metaphysical theory can never become well confirmed by experience, and so can never become knowledge. Thus, although metaphysics is meaningful and may be heuristically valuable for science in some cases, we can never lay claim to metaphysical knowledge. The realm of metaphysics is that of belief which does not have the status of knowledge.

Notes

Chapter 1 Some Historical Background

1   There is no doubt of the influence of Russell on Keynes (and indeed, vice versa) as regards issues of probability and induction. Thus Russell writes in the Preface to his *Problems of Philosophy*: 'I have derived valuable assistance from unpublished writings of... Mr J. M. Keynes... as regards probability and induction' (1912, p. v); while Keynes says in the Preface to his *Treatise on Probability*: 'It may be perceived that I have been influenced by W. E. Johnson, G. E. Moore, and Bertrand Russell, that is to say by Cambridge, which, with great debts to the writers of Continental Europe, yet continues in direct succession the English tradition of Locke and Berkeley and Hume, of Mill and Sidgwick' (1921, p. v). One point to note here is that Keynes mentions not only Bertrand Russell but also G. E. Moore. Recent research on the history of ideas has shown that this acknowledgement of Moore's influence is highly appropriate. In 1903 Moore published his famous treatise on ethics: *Principia ethica*. Keynes was most impressed with this work, but felt the need to alter and develop its ideas in some respects. This interest in ethics led him to an interest in probability. Thus Keynes's concern with the problems of probability and induction arose from the conjunction of two separate influences. One was the problem area of inductivism and scientific method, which was being investigated at the time by Russell and which is our concern in this book. The other was the problem area of ethics which was being studied by Moore. In this book I will not discuss Moore's views on ethics and their connection with questions of probability and induction, but rather refer the interested reader to Skidelsky, 1983; Bateman, 1988; and a forthcoming book by Davis on *Keynes's Philosophical Development*.

2   For further historical details concerning the relationship between Hume and Price, see Gillies, 1987.

3   For information about the early history of mathematical probability, see David's admirable 1962 book, which contains an English translation of the Fermat–Pascal correspondence of 1654 as appendix 4.

4   Those who would like to pursue the matter further can find an introduction to the work of Keynes, Ramsey, and de Finetti, as well as an account of some of the main criticisms of Bayesianism, in Gillies, 1988.

5   This account is based mainly on Frank, 1941, and Gadol (ed.), 1982. In the Gadol collection, Gadol's own essay and the memoirs of A. J. Ayer, Herbert Feigl, and Karl Menger were particularly useful. Some additional recollections by Karl Menger are to be found in Menger, 1980. The manifesto of the Vienna Circle (Neurath *et al.*, 1929) also contains some valuable historical details.

6   The exact contribution of Lorentz, Poincaré, and Einstein to the development of special relativity is a matter of dispute among experts in the history of physics. We can illustrate this by a brief account of the views of three scholars (Jerzy Giedymin, Arthur I. Miller, and Elie Zahar) who all produced carefully researched books on the subject in the 1980s. Miller (1981 and 1984, ch. 1 and 3) gives almost all the credit for the discovery of the special theory of relativity to Einstein. His view is that the work of Lorentz and Poincaré, though it had some points in common with that of Einstein, was lacking in the crucial insights and innovations which constitute special relativity. Thus, he says: 'Lorentz and Poincaré considered the Lorentz transformations as a mathematical device for deducing a principle of relativity for Lorentz's theory of the electron. . . . Einstein . . . considered the relativistic transformations to have a much deeper connotation' (1981, p. 217). And again: 'Poincaré deserved credit for having also obtained in 1905 the correct velocity addition law and the correct transformation equation for the charge density; but the similarity was only mathematical because the conceptual frameworks of Einstein and Poincaré were entirely dissimilar' (p. 325). This is why Miller speaks of *Albert Einstein's* Special Theory of Relativity.

     An entirely different view is expressed by Giedymin (1982, ch. 5). Giedymin summarizes his position on the question as follows: 'Critics of Edmund Whittaker's account of the discovery of special relativity have been unanimously defending the traditional discovery-by-one-man view . . . My own view on the matter . . . in terms of simultaneous discovery by Einstein, Lorentz and Poincaré, is in conformity with Fleck's and Robert Merton's account of scientific discoveries' (1986, p. 214, n. 14). Zahar (1989) analyses Einstein's discovery of special relativity in ch. 3, but then gives ch. 5 the title 'Poincaré's independent discovery of the relativity principle'.

     It should be stressed that the issue here is not simply one of priorities, but involves interesting and important philosophical questions. It is not easy to decide, for example, whether Poincaré's theory should be judged the same as, or different from, Einstein's. As Miller points out, the two theories share the same mathematics, but differ conceptually. Are these conceptual differences sufficient to make them different physical theories? Further, there is the general question of whether discovery in science is an individual act or a social process. We shall not be able to pursue these fascinating problems further in this book, although the interested reader is recommended to study the important works just cited. I will, however, say a little more about Poincaré's contributions to physics in Part II, where Poincaré's philosophy of science will be discussed.

7   Tu proverai sì come sa di sale
           Lo pane altrui, e com'è duro calle
           Lo scendere e il salir per l'altrui scale.
           Dante, *La Divina Commedia, Paradiso, xvii,* 58

('You shall make trial of how salt is the taste of another's bread, and how hard is the path of going up and down another's stair.')

Chapter 2   Popper's Critique of Inductivism

1   For those interested in studying Kepler's scientific work in more detail, the following accounts are recommended: Dreyer, 1906, pp. 372–412; Koestler, 1959, pp. 225–427; Koyré, 1961, pp. 117–464.
2   The historical details in this section and the next are largely based on Hare, 1970, and Macfarlane, 1984. I would like to express special thanks to my colleague Melvin Earles, who not only drew my attention to these two excellent books, but even lent me his own copies of them, as well as reading through my account and offering useful comments and corrections.
3   Peter Mitchell was awarded the Nobel Prize for Chemistry in 1978 for the chemiosmotic hypothesis, the first version of which he had proposed in 1961. He was thus an author with personal experience of scientific discovery. His 1989 paper on philosophy of science has so far appeared only in Japanese. I am most grateful to him for giving me a photocopy of an English version of the paper, from which the quotations in the text are taken.
4   The idea of this section came from some very stimulating conversations with Joseph D. Robinson, Professor of Pharmacology at SUNY Health Science Center, Syracuse, when we met at the Glynn Silver Jubilee Meeting in October 1990. Robinson pointed out to me that most drugs are discovered by screening, and argued that this raised severe difficulties for a Popperian 'conjectures and refutations' account of scientific development. Later he was kind enough to send me material relating to this and other questions in philosophy of science. This included the following quotation from Goth, 1970, p. 36: 'Most new drugs are discovered today by screening', on which Joseph Robinson commented that what this does not emphasize is that for whatever purpose a compound is synthesized, it is put through *all* the screens to see what 'other' activities it may have. This indicates that the discovery

of the sulphonamide drugs is quite typical of drug discovery in general, so that the analysis of this case is very important for the study of scientific method.

5   The following account of the 'dye heuristic' used by Domagk was suggested to me by Melvin Earles, who also lent me his copy of Ehrlich, 1906. Some comments from Professor W. C. Bowman of the Department of Physiology and Pharmacology at the University of Strathclyde were also useful.

Chapter 3   Duhem's Critique of Inductivism

1   It might be objected that the concepts of force and mass were not entirely novel in Newton's day. However, in my 1972 article I argue on historical grounds that 'to a first approximation at least we can say that the concept of mass as distinct from weight is original to Newton' (pp. 9–10), and that 'Newton's *quantitative* notion of *dynamical* force was indeed original to him' (p. 10). Through an analysis of this particular example, this article develops a general theory of conceptual innovation in the exact sciences. It contains a rather more technical account of the relationship between Newton's theory and Kepler's laws.

2   My account of Duhem's life and work is based on the foreword by Prince Louis de Broglie to the English edition of Duhem's *The Aim and Structure of Physical Theory*, the introductory essay by Stanley L. Jaki to the English edition of Duhem's *To Save the Phenomena*, and Jaki's own 1984 biography of Duhem. The books of Brenner (1990a) and Martin (1991) have also been useful. For Poincaré, I have used Bell's *Men of Mathematics* (1987), ch. 28, and *The Times*'s obituary of Poincaré, a photocopy of which was kindly sent to me by Jerzy Giedymin.

3   This view is supported by Maiocchi in his interesting 1990 article entitled 'Pierre Duhem's *The Aim and Structure of Physical Theory*: A Book against Conventionalism'.

4   Duhem writes; 'There is a general method of deduction; Aristotle has formulated its laws for all time (*pour toujours*)' (1915, p. 58). In fact, by 1915 the major logical writings of Frege, Peano, and Russell had all been published, and had clearly superseded Aristotelian logic. It is remarkable that Duhem seems to have known nothing of these developments.

5   Those interested in a further examination of Simon's work are recommended to look at *International Studies in the Philosophy of Science*, **6**, no. 1 (1992). This is a special issue of the journal devoted to the question. It contains an article by Simon, a series of papers commenting on this article (including an earlier version of the material in this section), and Simon's reply to his critics.

6   For further details about this approach, see the papers in Muggleton (ed.), 1992.

Chapter 4   Poincaré's Conventionalism of 1902

1   A good account of these experiments is to be found in Miller, 1981, pp. 47–54 and pp. 61–7.
2   The reference is to Le Roy whose paper 'Science et philosophie', *Revue de métaphysique et de morale*, 1901, is cited by Poincaré on p. xxiv of his preface to *Science and Hypothesis*.
3   For an excellent account of the Poincaré–Russell dispute, see J. Vuillemin's preface to the 1968 Flammarion edition of Poincaré 1902, esp. sec. 3 and 4, pp. 10–14.

Chapter 5   The Duhem Thesis and the Quine Thesis

1   1 degree = 60', and 1' = 60". So 52' is slightly less than a degree.
2   Einstein may have been influenced by Duhem, however, as is suggested by Howard in his interesting 1990 article. Howard shows that Einstein was on very friendly terms with Friedrich Adler, who prepared the first German translation of the *Aim and Structure of Physical Theory*, which appeared in 1908. From the autumn of 1909, Einstein and his wife rented an apartment in Zurich just immediately upstairs from the Adlers, and Einstein and Adler would meet frequently to discuss philosophy and physics. So probably Einstein had read *Aim and Structure* by the end of 1909 at the latest.
3   Or rather, misquotes. Duhem writes: 'raisons que la raison ne connaît pas' (1904–5, French edn, p. 330), whereas Pascal's original *pensée* was 'Le coeur a ses raisons que la raison ne connaît point.' Giving quotations which are slightly wrong is often a sign of great familiarity with a particular author.
4   Vuillemin (1979) and Ariew (1984) give valuable discussions of the differences between the Duhem thesis and the Quine thesis. I found these articles very helpful when forming my own views on the subject.
5   Quine's views have altered over the years, but here we will discuss only the position found in his 1951 article.
6   It is possible, however, to use arguments not involving the Quine thesis against the analytic/synthetic distinction. I give two such arguments against the distinction, the argument from justification and the argument from truth, in my 1985 article.

Chapter 6   Protocol Sentences

1   This seems to me rather unfair to Neurath, who also correctly emphasizes the need for science to have inter-subjective observation statements, and introduces the very important comparison between scientists developing science and sailors rebuilding a ship at sea.

Chapter 7   Is Observation Theory-Laden?

1   See Gregory, 1981, pp. 362–7. My discussion in this section is based largely on Gregory, 1970 and 1981.
2   I would like to thank Richard Gregory for kindly sending me the photograph used in Plate 4 and giving me permission to use it in this book. Richard Gregory also pointed out in conversation that some depth illusions can be produced by unusual lighting. This cannot be the case for the illusion shown in Plate 4, since standard lighting from above was used.

Chapter 8   Is Metaphysics Meaningless?

1   There are a number of excellent books about Wittgenstein's life. The most recent and comprehensive is Ray Monk's *Ludwig Wittgenstein. The Duty of Genius* (1990). This is enthralling to read, and I have taken from Monk the idea that Wittgenstein was profoundly religious in an unorthodox fashion and that this religious tendency is most important for understanding Wittgenstein's philosophy. For the period up to 1921, I have found Brian McGuinness's perceptive *Wittgenstein: A Life*. Vol. I: *Young Ludwig (1889–1921)* (1988) most useful. McGuinness has some valuable psychological analyses of Wittgenstein's relationships with his family and with Bertrand Russell. W. W. Bartley III's *Wittgenstein* (1973) contains some good insights; but unfortunately, some of Bartley's statements, particularly regarding Wittgenstein's homosexuality, are based on sources which Bartley refused to reveal (for a discussion, see appendix to Monk, 1990). I have found Bartley both reliable and helpful, however, on Wittgenstein's period as a village schoolmaster. Indeed, Bartley did some important pioneering investigations of this phase of Wittgenstein's life. The memoirs of Wittgenstein by Russell in his *Autobiography*. vol. 2 (1968) and by Malcolm in his *Ludwig Wittgenstein. A Memoir* (1958) are also important sources.

Chapter 10   Falsificationism in the Light of the Duhem–Quine Thesis

1   I introduced the principle of explanatory surplus in a paper read at a meeting of the British Society for the Philosophy of Science on 27 April 1987. I presented a modified version of this paper, with the title 'Non-Bayesian Confirmation Theory, and the Principle of Explanatory Surplus', to a biennial meeting of the Philosophy of Science Association in Chicago on 29 October 1988; this was subsequently published in the *Proceedings* (Gillies, 1989). Here, two sub-cases of the principle are considered, concerned with (a) number of parameters and (b) number of

theoretical assumptions. I subsequently discovered that principles very similar to both these cases had been introduced in the years 1986–8.

Thus, Mulaik (1986, p. 329) introduces an *index of parsimony*, which is closely related to case (a) above. The concept of parsimony is further discussed in Mulaik *et al.*, 1989, pp. 436–9. In particular, Mulaik relates parsimony to the degrees of freedom of a statistical test, something which is not done in my own 1989 paper.

In section 5.5.3, *Simplicity in PI*, Thagard, 1988, pp. 89–91, introduces a simplicity measure which is closely related to case (b) above. It is this case which will be dealt with in the present chapter.

While something like the principle of explanatory surplus has been a subject of interest in the last few years, research would probably reveal earlier formulations of much the same idea. One example which I came across occurs in one of Frege's early papers on logic, where Frege proposes what is in effect a special case of the principle of explanatory surplus. He says: 'The value of an explanation can be directly measured by this condensation and simplification: it is zero if the number of assumptions is as great as the number of facts to be explained' (*c*. 1880/1, p. 36).

Moreover, the principle is perhaps implicit in the account which Duhem gives of a physical theory in a passage already quoted in 9.3: 'A physical theory . . . is a system of mathematical propositions deduced from a small number of principles, which aim to represent as simply, as completely, and as exactly as possible a set of experimental laws' (1904–5, p. 19). But Duhem does not carry the idea further, since he nowhere tries either to develop a theory of confirmation or to formulate a criterion for demarcating science from metaphysics.

2   In preparing my account of this important scientific development, I have made particular use of Max Born's *Atomic Physics* (1935), chapters 4 and 8, and of Bernard Cohen's *Revolution in Science* (1985), chapter 27.

3   See chapter 2, note 3. Peter Mitchell had a great (though critical) admiration for Popper's philosophy of science, and he consciously tried in his own scientific research to apply some of Popper's methodological principles.

4   My information is derived from Tim Jeal's excellent 1973 biography of Livingstone, particularly cf. pp. 50 and 69.

5   For a valuable discussion of some of these ethical questions, as well as further references, see Botros, 1990.

References

In general, works are cited by their date of first publication, but the exact edition from which quotations are taken is also specified, and its date given if different from the first edition. For example, if the source of a quotation is given as Bacon, 1620, p. 245, the list of references shows that the quotation is from Bacon's *Novum Organum*, which was first published in 1620, but that the page number refers to an English translation published in 1905. Sometimes a date will be to the second or a subsequent edition if this differs significantly from the first and is more appropriate in the context. Regarding Kant's *Critique of Pure Reason*, the first edition of 1781 and the second edition of 1787 are both important, so that it is referred to as Kant, 1781/7. Duhem's *The Aim and Structure of Physical Theory* was first published as a series of papers, some of which appeared in 1904, the rest in 1905. It is therefore referred to as Duhem, 1904–5. Occasionally, if a work was first published long after its completion, it is referred to as, for example, *c*.1880/1, where the *c*. indicates that the work was completed about the date shown.

Anscombe, G. E. M. 1959: *An Introduction to Wittgenstein's* Tractatus. Hutchinson University Library.

Ariew, R. 1984: The Duhem Thesis. *British Journal for the Philosophy of Science*, **35**, 313–25.

Ayer, A. J. 1946: *Language, Truth and Logic*. 2nd edn, Gollancz; 15th impression, 1962.

Bacon, F. 1620: *Novum Organum*. English translation in R. L. Ellis and J. Spedding (eds), *The Philosophical Works of Francis Bacon*, Routledge, 1905, 212–387.

Bartley, W. W. III 1973: *Wittgenstein*. Quartet Books, 1974.

Bateman, B. W. 1988: G. E. Moore and J. M. Keynes: A Missing Chapter in the History of the Expected Utility Model. *American Economic Review*, **78**, 1098–1106.

Bell, E. T. 1937: *Men of Mathematics*. Pelican edn, 1965.

Berlin, I. 1939: Verifiability in Principle. *Proceedings of the Aristotelian Society*, **39**, 225–48. (This paper is listed as 'Verifiability in Principle' in the table of contents of the volume, but the paper itself is headed 'Verification'.)

Born, M. 1935: *Atomic Physics*. 7th edn, Blackie, 1962.

Botros, S. 1990: Equipoise, Consent and the Ethics of Randomised Clinical Trials. In P. Byrne (ed.), *Ethics and Law in Health Care and Research*, Wiley, 9–24.

Brenner, A. A. 1990a: *Duhem: Science, réalité et apparence*. Vrin.

Brenner, A. A. 1990b: Holism a Century Ago: The Elaboration of Duhem's Thesis. *Synthese*, **83**, 325–35.

Carnap, R. 1931: The Logicist Foundations of Mathematics. Reprinted in English translation in P. Benacerraf and H. Putnam (eds), *Philosophy of Mathematics. Selected Readings*, 2nd edn, Cambridge University Press, 1983, 41–52.

Carnap, R. 1932: The Elimination of Metaphysics through Logical Analysis of Language. Reprinted in English translation in A. J. Ayer (ed.), *Logical Positivism*, Free Press, 1959, 60–81.

Carnap, R. 1932/3: Psychology in Physical Language. Reprinted in English translation in A. J. Ayer (ed.), *Logical Positivism*, Free Press, 1959, 165–98.

Carnap, R. 1950: *Logical Foundations of Probability*. University of Chicago Press. 2nd edn, 1963.

Carnap, R. 1963: *Intellectual Autobiography*. In P. A. Schilpp (ed.), *The Philosophy of Rudolf Carnap*, Library of Living Philosophers, Open Court, 3–84.

Cohen, I. B. 1985: *Revolution in Science*. Harvard University Press.

Crowe, M. J. 1990: Duhem and the History and Philosophy of Mathematics. *Synthese*, **83**, 431–47.

David, F. N. 1962: *Games, Gods and Gambling*. Hafner.

De Oliveira, M. B. 1978: Popper's Two Problems of Demarcation. *Proceedings of the Third International Wittgenstein Symposium*, 402–5.

Dreyer, J. L. E. 1906: *A History of Astronomy from Thales to Kepler*. Dover edn, 1953.

Duhem, P. 1904–5: *The Aim and Structure of Physical Theory*. English translation by Philip P. Wiener of the 2nd French edn of 1914, Atheneum, 1962. French edn, Vrin, 1989.

Duhem, P. 1905: Physics of a Believer. Reprinted as an appendix to Duhem, 1904–5, pp. 273–311.

Duhem, P. 1908a: *To Save the Phenomena*. English translation with an introductory essay by Stanley L. Jaki, University of Chicago, 1969.

Duhem, P. 1908b: The Value of Science. Reprinted as an appendix to Duhem, 1904–5, pp. 312–35.

Duhem, P. 1915: *La Science allemande*. A. Hermann et Fils.

Ehrlich, P. 1906: Address Delivered at the Dedication of the Georg-Speyer-Haus. Reprinted in English translation in L. Shuster (ed.), *Readings in Pharmacology*, Churchill, 1962, 233–43.

Einstein, A. 1905: Zur Electrodynamik bewegter Körper. *Annalen der*

Physik, **17**, 891–921. English translation as 'On the Electrodynamics of Moving Bodies', in Miller, 1981, pp. 392–415.

Engels, F. 1883: Speech at the Graveside of Karl Marx. In *Karl Marx and Frederick Engels: Selected Works*, Lawrence and Wishart, 1968, 429–30.

Evans, B. and Waites, B. 1981: *IQ and Mental Testing. An Unnatural Science and its Social History*. Macmillan.

Feyerabend, P. 1975: *Against Method*. Verso, 1984.

Fleming, A. 1929: On the Antibacterial Action of Cultures of a Penicillium, with Special Reference to their Use in the Isolation of *B. Influenzae*. *British Journal of Experimental Pathology*, **10**, 226–36.

Frank, P. 1941: *Modern Science and its Philosophy*. Paperback edn, Collier Books, 1961.

Frege, G. *c*.1880/1: Boole's Logical Calculus and the Concept-script. In Frege, *Posthumous Writings*, Blackwell, 1979, 9–46.

Freud, S. 1917: One of the Difficulties of Psycho-Analysis. In Freud, *Collected Papers*, vol. 4, Hogarth Press, 1957, 347–56.

Gadol, E. (ed.) 1982: *Rationality and Science. A Memorial Volume for Moritz Schlick in Celebration of the Centennial of his Birth*. Springer Verlag.

Galileo 1610: *The Starry Messenger*. Reprinted in *Discoveries and Opinions of Galileo*, English translation with an introduction and notes by S. Drake, Doubleday Anchor, 1957, 21–58.

Giedymin, J. 1982: *Science and Convention. Essays on Henri Poincaré's Philosophy of Science and the Conventionalist Tradition*. Pergamon.

Giedymin, J. 1986: Polish Philosophy in the Inter-War Period and Ludwik Fleck's Theory of Thought-Styles and Thought-Collectives. In R. S. Cohen and T. Schnelle (eds), *Cognition and Fact – Materials on Ludwik Fleck*, Reidel, 179–215.

Giedymin, J. 1991: Geometrical and Physical Conventionalism of Henri Poincaré in Epistemological Formulation. *Studies in History and Philosophy of Science*, **22**, 1–22.

Gillies, D. A. 1971: A Falsifying Rule for Probability Statements. *British Journal for the Philosophy of Science*, **22**, 231–61.

Gillies, D. A. 1972: Operationalism. *Synthese*, **25**, 1–24.

Gillies, D. A. 1973: *An Objective Theory of Probability*. Methuen.

Gillies, D. A. 1982: *Frege, Dedekind, and Peano on the Foundations of Arithmetic*. Van Gorcum.

Gillies, D. A. 1985: The Analytic/Synthetic Problem. *Ratio*, **27**, 149–59.

Gillies, D. A. 1987: Was Bayes a Bayesian? *Historia Mathematica*, **14**, 325–46.

Gillies, D. A. 1988: Induction and Probability. In G. H. R. Parkinson (ed.), *An Encyclopaedia of Philosophy*, Routledge, 179–204.

Gillies, D. A. 1989: Non-Bayesian Confirmation Theory, and the Principle of Explanatory Surplus. In A. Fine and J. Leplin (eds), *Philosophy of Science Association, 1988*, vol. 2, 373–80.

Gillies, D. A. 1990: Bayesianism versus Falsificationism. *Ratio*, **3**, 82–98.

Goth, A. 1970: *Medical Pharmacology*. C. V. Mosby.

Gregory, R. L. 1970: *The Intelligent Eye.* George Weidenfeld and Nicolson.

Gregory, R. L. 1981: *Mind in Science.* Penguin edn, 1981.

Hanson, N. R. 1958: *Patterns of Discovery.* Cambridge University Press, 1965.

Hare, R. 1970: *The Birth of Penicillin and the Disarming of Microbes.* Allen & Unwin.

Howard, D. 1984: Realism and Conventionalism in Einstein's Philosophy of Science: The Einstein–Schlick Correspondence. In P. Weingartner and C. Puhringer (eds), *Philosophy of Science. History of Science. A Selection of Contributed Papers of the Seventh International Congress of Logic, Methodology and Philosophy of Science, Salzburg, 1983,* Anton Hain-Mesenheim/Glan, 616–29.

Howard, D. 1990: Einstein and Duhem. *Synthese,* **83,** 363–84.

Howson, C. and Urbach, P. 1989: *Scientific Reasoning. The Bayesian Approach.* Open Court.

Hume, D. 1748: *Enquiry concerning the Human Understanding.* Selby-Bigge edn, Oxford, 1963.

Jaki, S. L. 1984: *Uneasy Genius: The Life and Work of Pierre Duhem.* Martinus Nijhoff.

Jaki, S. L. (ed.) 1987: *Pierre Duhem. Prémices philosophiques.* E. J. Brill.

Janik, A. and Toulmin, S. 1973: *Wittgenstein's Vienna.* Simon and Schuster.

Jeal, T. 1973: *Livingstone.* Penguin edn, 1985.

Kant, I. 1781/7: *Critique of Pure Reason.* English translation by Norman Kemp Smith, Macmillan, 1958.

Kant, I. 1783: *Prolegomena to any Future Metaphysics that will be able to Present itself as a Science.* English translation by Peter G. Lucas, Manchester University Press, 1959.

Keynes, J. M. 1921: *A Treatise on Probability.* Macmillan, 1963.

Koestler, A. 1959: *The Sleepwalkers. A History of Man's Changing Vision of the Universe.* Pelican edn, 1968.

Koyré, A. 1961: *The Astronomical Revolution. Copernicus – Kepler – Borelli.* English translation by R. E. W. Maddison, Methuen, 1973.

Kuhn, T. S. 1962: *The Structure of Scientific Revolutions.* University of Chicago Press.

Kuhn, T. S. 1970: Logic of Discovery or Psychology of Research. In I. Lakatos and A. Musgrave (eds), *Criticism and the Growth of Knowledge,* Cambridge University Press, 1–23.

Lakatos, I. 1968: Changes in the Problem of Inductive Logic. Reprinted in J. Worrall and G. Currie (eds), *Imre Lakatos: Philosophical Papers,* vol. 2, Cambridge University Press, 1978, 128–200.

Langley, P.; Simon, H. A.; Bradshaw, G. L.; and Zytkow, J. M. 1987: *Scientific Discovery. Computational Explorations of the Creative Processes.* MIT Press.

Lenin, V. I. 1908: *Materialism and Empirio-criticism. Critical Comments on a Reactionary Philosophy.* Progress Publishers, 1970.

Macfarlane, G. 1984: *Alexander Fleming. The Man and the Myth*. Chatto & Windus.

McGuinness, B. F. (ed.) 1967: *Ludwig Wittgenstein und der Wiener Kreis*. Blackwell.

McGuinness, B. F. 1988: *Wittgenstein: A Life*. vol. 1: *Young Ludwig (1889–1921)*. Penguin edn, 1990.

Magee, B. (ed.) 1971: *Modern British Philosophy*. Secker & Warburg.

Maiocchi, R. 1990: Pierre Duhem's *The Aim and Structure of Physical Theory*: A Book against Conventionalism. *Synthese*, **83**, 385–400.

Malcolm, N. 1958: *Ludwig Wittgenstein: A Memoir*. Oxford University Press, 1962.

Martin, R. N. D. 1991: *Pierre Duhem. Philosophy and History in the Work of a Believing Physicist*. Open Court.

Menger, K. 1980: Introduction to Hans Hahn, *Empiricism, Logic, and Mathematics: Philosophical Papers*, ed. Brian McGuinness, Reidel, ix–xviii.

Menger, K. 1982: Memories of Moritz Schlick. In Gadol (ed.), 1982, pp. 83–103.

Miller, A. I. 1981: *Albert Einstein's Special Theory of Relativity. Emergence (1905) and Early Interpretaion (1905–1911)*. Addison-Wesley.

Miller, A. I. 1984: *Imagery in Scientific Thought*. MIT Press, 1986.

Mitchell, P. D. 1989: Aspects of Chemical Philosophy. Science as a Pursuit of Humanity. *Kagaku to Kogyo*, **42**, 60–9. (This published version is in Japanese. Quotations are taken from a photocopy of an English version of the paper given to Donald Gillies by Peter Mitchell.)

Monk, R. 1990: *Ludwig Wittgenstein. The Duty of Genius*. Jonathan Cape.

Muggleton, S. (ed.) 1992: *Inductive Logic Programming*. Academic Press.

Mulaik, S. A. 1986: Towards a Synthesis of Deterministic and Probabilistic Formulations of Causal Relations by the Functional Relation Concept. *Philosophy of Science*, **53**, 313–32.

Mulaik, S. A.; James, L. R.; Van Alstine, J.; Bennett, N.; Lind, S.; and Stilwell, C. D. 1989: Evaluation of Goodness-of-Fit Indices for Structural Equation Models. *Psychological Bulletin*, **105**, 430–45.

Neurath, O. *et al.* 1929: *The Scientific Conception of the World. The Vienna Circle*. English translation, Reidel, 1973.

Neurath, O. 1932/3: Protocol Sentences. Reprinted in English translation in A. J. Ayer (ed.), *Logical Positivism*, Free Press, 1959, 199–208.

Newton, I. 1687: *Philosophiae Naturalis Principia Mathematica*. Andrew Motte's English translation of 1729, revised by Florian Cajori, University of California Press, 1960.

Pitcher, G. 1964: *The Philosophy of Wittgenstein*. Prentice-Hall.

Poincaré, H. 1902: *Science and Hypothesis*. English translation, Dover, 1952. French edn, Flammarion, 1968.

Poincaré, H. 1904: L'État actuel et l'avenir de la physique mathématique. Lecture delivered 24 Sept. 1904 to the International Congress of Arts and Science, St Louis, Missouri, and published in *Bulletin des sciences mathématiques*, **28**, 302–24. Reprinted in Poincaré, 1905, pp. 91–111.

Poincaré, H. 1905: *The Value of Science*. English translation, Dover, 1958.

Poincaré, H. 1906a: Reply to Russell's review of *Science and Hypothesis*. *Mind*, n.s. **15**, 141–3.

Poincaré, H. 1906b: Sur la dynamique de l'électron. *Rendiconti del circolo matematico di Palermo*, **21**, 129–75. Reprinted in Poincaré, *Oeuvres*, vol. 9, 494–550.

Poincaré, H. 1908: *Science and Method*. English translation, Dover.

Popper, K. R. 1934: *The Logic of Scientific Discovery*. 6th (rev.) impression of the 1959 English translation, Hutchinson, 1972.

Popper, K. R. 1963: *Conjectures and Refutations. The Growth of Scientific Knowledge*. Routledge & Kegan Paul.

Popper, K. R. 1976: *Unended Quest. An Intellectual Autobiography*. Fontana/Collins.

Popper, K. R. 1982a: *The Open Universe. An Argument for Indeterminism*. Hutchinson.

Popper, K. R. 1982b: *Quantum Theory and the Schism in Physics*. Hutchinson.

Popper, K. R. 1983: *Realism and the Aim of Science*. Hutchinson.

Putnam, H. 1975: *Philosophical Papers*, vol. 1. Cambridge University Press.

Quine, W. V. O. 1951: Two Dogmas of Empiricism. Reprinted in *From a Logical Point of View*, 2nd rev. edn, Harper Torchbooks, 1961, 20–46.

Ramsey, F. P. 1926: Truth and Probability. Reprinted in H. E. Kyburg and H. E. Smokler (eds), *Studies in Subjective Probability*, Wiley, 1964, 61–92.

Russell, B. 1897: *An Essay on the Foundations of Geometry*. Cambridge University Press.

Russell, B. 1905: Review of *Science and Hypothesis* by H. Poincaré. *Mind*, n.s. **14**, 412–18.

Russell, B. 1912: *The Problems of Philosophy*. Williams and Norgate, n.d.

Russell, B. 1968: *Autobiography*, vol. 2. Allen & Unwin.

Simon, H. A. 1992: Scientific Discovery as Problem Solving. *International Studies in the Philosophy of Science*, **6(1)**, 3–14.

Skidelsky, R. 1983: *John Maynard Keynes*. vol. 1. *Hopes Betrayed 1883–1920*. Macmillan.

Thagard, P. 1988: *Computational Philosophy of Science*. MIT Press.

Tolstoy, L. 1879: A Confession. In *A Confession, The Gospel in Brief and What I Believe*, English translation with an introduction by Aylmer Maude, Oxford University Press, 1971.

Von Wright, G. H. 1958: Biographical Sketch. Printed with *Ludwig Wittgenstein: A Memoir* by N. Malcolm, Oxford University Press, 1–22.

Vuillemin, J. 1968: *Préface* to H. Poincaré, *La Science et l'hypothèse*, Flammarion, 1968, 7–19.

Vuillemin, J. 1979: On Duhem's and Quine's Theses. *Grazer Philosophische Studien*, **9**, 69–96. Quotations are from reprint in L. E. Hahn and

P. A. Schilpp (eds), *The Philosophy of W. V. Quine*, Library of Living Philosophers, Open Court, 1986, 595–618.

Wittgenstein, L. 1921: *Tractatus Logico-Philosophicus*. English translation by D. F. Pears and B. F. McGuinness, Routledge & Kegan Paul, 1963.

Wittgenstein, L. 1953: *Philosophical Investigations*. English translation by G. E. M. Anscombe, Blackwell, 1963.

Wittgenstein, L. 1956: *Remarks on the Foundations of Mathematics*. Blackwell, 1967.

Zahar, E. 1989: *Einstein's Revolution. A Study in Heuristic*. Open Court.

Index

Abraham, M. 114
Adams, J. C. 100, 111
Adler, A. 206, 218, 220–1
Adler, F. 235
agar 40
Albumasar 196
Ames, A. 143–4
Ames room 143–4
Ampère, A. M. 107
analytic 76–7, 108–10, 235
Anscombe, G. E. M. 167
antiseptic(s) v, 40, 43–7
aphelion 37–8
Aquinas, St Thomas 175, 203
Arago, F. 102
Ariew, R. 235
Aristotle 4, 37, 91, 115, 175, 196, 204
artificial intelligence vii, 32, 68–9, 72, 115
astrology 154–5, 196
atomism 189–90, 193, 197–200, 204, 215, 224
Austin, J. L. 22
Ayer, A. J. 25, 213, 232
Ayer example 213–15, 217

Bacon, F. 3–5, 7, 15–16, 26, 32–3, 49–50, 55, 101, 189
Bartley, W. W. 161–5, 187, 236
basic statements viii, 124–30
Bateman, B. W. 231
Bayes, T. 14–15
Bayesian(s) 14–17, 31–4, 95, 209

Bell, E. T. 234
Beltrami, E. 83
Berkeley, G. 119, 231
Berlin, I. 213
Bernard, C. 110
Berthelot, P. E. M. 63–4
Bessel, F. W. 80
Biot, J. B. 106
black body radiation 223, 226
Bohr, N. 61, 194, 199
Boltzmann, L. 64
Bolyai, J. 80–2
Bolyai, W. 80
Bolyai–Lobachevsky geometry 78–80
Boole, G. 57
Born, M. xiv, 135, 225, 237
Botros, S. 237
Bowman, W. C. 234
Boyle, R. 69
Bradshaw, G. L. 69, 71
Brahe, T. 7, 27, 38–9, 70
Brahms, J. 157
Braithwaite, R. 23
breast cancer 229
Brenner, A. A. xiii, 61, 63, 234
Britton, K. 184
Broglie, L. de 234
Brouwer, L. E. J. 115, 163
Brownian motion 225

Cambridge school vii, 3, 8, 11, 16–17, 34
Carnap, R. viii, 13, 16–18, 31–2,

67–8, 109, 120–5, 134, 146–7,
156, 172–6, 178
Casals, P.   157
cathode rays   93–4
Cauchy, A.-L.   195
chemiosmotic hypothesis   xiv, 201,
227, 233
Cohen, I. B.   227, 237
computer(s)   31–2, 68–72
concerto for the left hand   158
confirmability   206, 210, 213–15
conjecture(s)   vii, 26, 29, 32–3,
35–6, 38–9, 45–9, 71, 205–6,
222, 233
connectives   170–1, 186
consciousness   148
conservation of mass   93
conventionalism   viii, 66–7, 73,
75, 90, 94–5, 97, 102, 104, 109,
211
Copernicus, N.   27, 48, 65, 70, 154
creationists   154
creative theorizing   48
creativity   30, 43, 45, 48–9, 72
Crelle, A.   81
Crowe, M. J.   113
crucial experiment   viii, 98, 101–2

Dalton, J.   190–1
Dante   24, 233
Darwin, C.   15
David, F. N.   231
Davis, J.   231
De Oliveira, M. B.   155
deferent   38, 70
demarcation   viii, 19, 21, 151,
153–6, 160, 164, 177–8, 180,
185, 205–6, 210, 212–14, 221,
228–30
demarcation criterion   ix, 177–8,
180, 193, 195, 205–6, 210,
212–14, 221, 228, 230, 237
Democritus   190
Descartes, R.   62, 67, 195, 197
Dirac, P.   20, 61, 134, 199
discovery(ies)   vii, viii, xi, xiii,
xiv, 4, 7, 30–2, 36, 46–8, 50,

57, 77, 100, 195, 198, 200,
233–4
displacement law (Wien's)   223–4
Domagk, G.   48–51
Dreyer, J. L. E.   233
duck-rabbit   140–2
Duhem, P.   vii–ix, xiii, 3, 25,
53–4, 58–67, 89, 94, 97–115,
132–3, 136–9, 189, 192–204,
206, 212, 219–20, 222–4,
234–5, 237
Duhem thesis   viii, 89, 98–100,
110–12, 115, 206, 235
Duhem–Quine thesis   viii, ix, 73,
98, 112, 115–16, 205, 210–12,
236
dye heuristic   51–2, 234

Earles, M. P.   xiii, 233–4
eclipse experiment   9, 20, 85
Ehrlich, P.   51, 234
Einstein, A.   xii, 9, 18, 20–2, 31,
60–1, 64, 66, 85, 88, 94, 100,
105, 107, 115, 185, 199–200,
218, 223, 225–8, 232, 235
elementary proposition(s)   166–8,
186
Engelmann, P.   163
Engels, F.   154
Epicurus   190
epicycle(s)   38–9, 70
Euclid   75, 77–80, 113
Evans, B.   155
existential statements   ix, 179,
205–6
experiments of fruit   5
experiments of light   5
explanatory surplus, principle of
ix, 214–20, 222, 225, 228, 236
Eysenck, H.   155

falsifiability   ix, 177–8, 180,
205–6, 208, 210–14, 221
falsifiability criterion   ix, 177–8,
180, 205, 211, 213
falsificationism   vii, ix, 26, 29, 36,
46–7, 98–9, 116, 205–6,

209–10, 221–3, 227–8, 236
Feigl, H.   17, 61, 177, 232
Fermat, P. de   16, 231
Feyerabend, P.   134
Finetti, B. de   16, 232
Fisher, R. A.   209
Fleck, L.   232
Fleming, A.   v, vii, xi, xiv, 36,
    39–48, 128, 130, 137
Florey, H. W.   46
formal logic   xi, 57, 67–8, 193
Foucault, L.   101–2, 106–7, 212
Fraenkel, A.   13
Frank, P.   17–18, 119, 232
Frege, G.   11–12, 57, 67, 76, 96,
    108–9, 115, 186, 193, 237
Fresnel, A.   106
Freud, S.   154
Fries, J. F.   124–6, 147–8
Fries's trilemma   124–6, 147

Gadol, E.   18, 232
Galen   196
Galileo   62, 64–5, 69–70, 134, 197,
    219
Gassendi, P.   195
Gauss, K. F.   80–2
Genesis   154
Gibbs, J. W.   64
Giedymin, J.   xiii, 97, 232, 234
Gillies, D. A.   iii, 209, 222, 231–2,
    236
Giorello, G.   xiv
Gödel, K.   13, 17, 67
good sense (le bon sens)   viii, 102,
    105–8, 112, 114–15, 199
Gosset, W. S. ('Student')   209
Goth, A.   233
Gregory, R. L.   xiv, xv, 140,
    142–5, 236

Hahn, H.   17–19, 156
Hanson. N. R.   142
Hare, R.   41, 43, 233
Hegel, G. W. F.   155
Heidegger, M.   172–4
Heisenberg, W.   20, 61, 199

Helmholtz, H. von   64, 107–8,
    140, 195
Hertz, H.   199, 225
heuristic(s)   51–2, 70–2, 191, 201,
    222, 225–7, 234
Hitler, A.   158, 174
holistic thesis   viii, 112–13, 115,
    137
Holmes, Sherlock   6
Hoüel, J.   82
Howard, D.   20, 235
Howson, C.   209
Hume, D.   7, 15, 21, 29, 57, 153,
    231
Huygens, C.   62, 102, 212

impersonal observation
    statements   128, 130–1, 149
induction   vii, 8–11, 14–15, 19,
    21, 26–8, 34, 39, 46–50, 54,
    56–7, 59, 62–3, 69, 71, 95–6,
    156, 231
    Baconian   vii, 48, 50, 69, 71
    conjectural   47, 62–3
    creative   vii, 39, 48
    mechanical   vii, 48–9
    principle of   vii, 8–11, 14, 27–8,
    34
    by simple enumeration   50
inductivism   vii, 1, 3, 5–6, 8,
    14–15, 17, 25–6, 31–3, 35–6,
    46–7, 53–6, 58, 60–3, 68–9,
    132, 231, 233–4
inferiority complex   206, 218,
    220–1
instruments   133–4, 137, 146
inter-subjective   120–1, 124–5,
    130, 136, 146–7, 235
IQ tests   155

Jaki, S. L.   61, 234
Janik, A.   168
Jeal, T.   237
Jeans, J.   224
Jeffreys, H.   14
Joachim, J.   157
Johnson, W. E.   14, 231

justification   vii, 30–1, 32, 34–5, 57

Kamminga, H.   xiii
Kant, I.   viii, 22, 28, 30, 75–7, 86–7, 95, 108–9, 148, 153
Kaufmann, W.   93–4
Kekulé, F. A.   30
Kepler, J.   vii, xi, 7, 27, 35–9, 46, 48, 58–60, 62, 69–72, 99, 197, 210, 216, 219–20, 233
Kepler's laws   vii, 58–60, 62–3, 216, 219–20, 234
Keynes, J. M.   14, 16, 231–2
Klein, K. F.   83
Klimt, G.   157
Koestler, A.   233
Koyré, A.   233
Kraft, V.   17
Kuhn, T. S.   38, 68

La Touche, C. J.   42
Lakatos, I.   57
Langley, P.   69, 71
language-game(s)   181–2, 184, 188
Laplace, P. S.   136, 191–2
Laue, M. von   114
Lavoisier, A.   93, 194
Lavoisier's principle   93
law of gravity (or gravitation)   vii, 9, 54–5, 57–60, 62–3, 99, 104–5, 197, 218–19
Le Roy, E.   94, 102, 235
Leibniz, G. W.   57, 155, 191
Lenard, P.   225–6
Lenin, V. I.   119
Leonardo da Vinci   64
Leucippus   190
Leverrier, U.   100, 111
Levi-Cività, T.   185
linguistic philosophy   22
Livingstone, D.   229, 237
Lobachevsky, N. I.   80–2, 89
Locke, J.   231
logical empiricism   11, 14, 18–19, 25
logicism   11–14, 19, 67, 76, 96

Lorentz, H.   20, 64, 66, 94, 200, 232
Lucretius   190
lysis   41, 128, 137
lysozyme   44–5

Macfarlane, G.   233
McGuinness, B. F.   19, 159, 164, 236
Mach, E.   64, 119, 211
machine learning   32, 69, 71
Magee, B.   181
Mahler, G.   157
Maiocchi, R.   234
Malcolm, N.   164, 167–8, 236
Malebranche, N. de   195
Martin, R. N. D.   xiii, 107, 175, 202, 234
Marx, K.   154
Marxism   24, 119, 154–5
Maxwell, J. C.   64, 107–8, 190–1, 195, 197, 199–200, 227
mechanical falsificationism   49–51, 71
Menger, K.   18–19, 156, 232
Merton, R.   232
metaphysical research programme   190
metaphysics   viii, ix, 19, 23, 120, 149, 151, 153, 155–6, 160, 165–6, 171–3, 175–8, 183, 185, 189–95, 197, 199–203, 205–6, 213, 215, 219, 222, 229–30, 236–7
Michelangelo   xii
Mill, J. S.   11, 231
Miller, A. I.   xiii, 200, 232, 235
Millikan, R. A.   227–8
mind–body problem   148
Minkowski, H.   114
Mises, R. von   22
Mitchell, P. D.   xiv, 47, 201, 227, 233, 237
modified falsificationism   67, 104, 223
Monk, R.   162, 236
Moore, G. E.   11, 163, 231

Morin, J. B.   196
Morrell, Lady Ottoline   159
Mozart, W. A.   xii
Muggleton, S.   71, 234
Mulaik, S. A.   237
mystical   174–6, 203

Nazi(s)   xii, 24
Necker, L. A.   140–2
Necker cube   140–2
Neptune   100
Neurath, O.   v, viii, 17–18, 20–1,
    24, 119–20, 122–9, 134,
    137–8, 146–7, 232, 235
Neurath's principle   viii, 137–8
Newcomb, S.   100
Newton, I.   vii, 9, 21, 29, 54–60,
    62–3, 75, 91, 93–4, 99–100,
    102, 104–5, 111–12, 197,
    211–12, 215, 218–19, 234
Newtonian mechanics   20–1, 29,
    54, 60–1, 90–4, 102, 105,
    186–7, 191–2, 206
Newtonian method   vii, 54–5, 61
Newton's first law of motion (or
    the principle of inertia)   58,
    90–1, 99, 103, 111, 211–13,
    215, 222
Neyman, J.   209
Neyman paradox   209
Nietzsche, F.   173–4
non-Euclidean geometry   viii, xi,
    11, 75, 77–83, 85–8, 113–14,
    199
normal science (or scientist)   38–9,
    70–1
Novum Organum   4–5, 15–16, 32,
    101

observation(s)   viii, xiv, 5–7, 11,
    14–16, 21, 29, 36, 38–9, 46–9,
    87, 95, 99, 103–6, 110, 112,
    114, 116–17, 119–20, 122,
    124–5, 128–9, 132–4, 136–44,
    146–9, 167, 169–70, 172–3,
    178–9, 210, 213–16, 221, 230,
    235–6

observation statement(s)   viii, 99,
    110, 119–20, 122, 124–5,
    128–9, 133, 137, 139, 142,
    146–9, 167, 169–70, 172–3,
    178–9, 210, 213–16, 221, 235
Ohm, G. S.   69
Ostwald, W.   64

parallel postulate   77–9, 81
parsimony   237
Pascal, B.   16, 107, 231, 235
Peano, G.   11, 67, 115, 193
Pearson, E. S.   209
Pearson, K.   209
penicillin   v, vii, xi, xiii, xiv, 36,
    39, 41–3, 45–7, 128
perihelion   37–8, 85, 100, 105
    of Mercury   85, 100, 105
Petri dishes   40
phagocytes   43–5, 47
phenomenalism   119
photoelectric effect   225–8
physicalism   119–20, 123, 134, 147
Pitcher, G.   184
Planck, M. K. L.   xii, 20, 218,
    223–4, 226
Planck's constant   224, 226, 228
Planck's radiation law   224, 228
planet(s)   vii, xi, 7, 27, 36
Plato   37, 155, 194, 230
Playfair, J.   78
Poincaré, H.   vii, viii, xiii, 20, 25,
    60–1, 63, 65–7, 75, 77, 85–97,
    102–3, 105, 107–8, 114, 211,
    232–5
Popper, K. R.   vii, viii, 3, 6–7,
    10–11, 21–3, 25–35, 38–9, 47,
    49, 53, 57, 60–2, 98–9, 116,
    124–32, 140, 146–9, 153, 155,
    172, 177–81, 189–92, 195,
    197–8, 205–8, 210–13, 215,
    220–2, 233, 237
Price, R.   14–15, 231
private language   122, 124, 147
probability(ies)   ix, 9–10, 14–16,
    21–2, 32–5, 95–6, 192, 205,
    207–9, 231

projective geometry 87
Prontosil rubrum 49–52
protocol sentence(s) (or protocol(s))
    viii, 119–24, 126, 128–30,
    137–8, 147, 149, 167, 235
Proust, M. xii
Pryce, D. M. 40
pseudo-science 154–6
psycho-analysis 154–5
psychologism 119–20, 123–6,
    130, 146
Ptolemy 196
Putnam, H. 75
Pythagorean(ism) 48, 70, 201

quantum theory (or mechanics) xi,
    20–2, 61–2, 115, 192,
    199–200, 218, 223, 226–8
Quine, W. V. O viii, 25, 67, 98,
    108–13, 115, 124, 189, 192–4,
    206, 235
Quine thesis viii, 98, 108, 110–12,
    115, 235

rain-making ceremony 229
Ramsey, F. P. 14, 16–17, 186, 232
Ravel, M. 158
Rayleigh, J. W. S. 224
Rayleigh–Jeans radiation law 224
refutation(s) vii, 26, 29, 32, 35–6,
    38–9, 46–7, 49, 71, 104, 205,
    222, 233
Regnault, J. B. 136
Reichenbach, H. 10, 17, 22
relativity 9, 20–1, 25, 29, 60–1,
    64, 66, 85–6, 88, 90, 94, 97,
    100, 105, 107, 114–15, 185,
    199–200, 225, 228, 232
religion ix, xii, 153–5, 174–6,
    179, 201–3
revolution in computing 68
revolution in physics vii, xiii,
    20–1, 54, 60–1, 68, 92–3, 105
Rey, A. 201
Ricci, G. C. 185
Riemann, B. 81–2, 88–9, 114

Riemannian geometry 78–81,
    83–6, 88, 90, 113
Robinson, J. D. xiv, 233
Rosser, J. 13
Rudolfine tables 39
Russell, B. vii, 3, 8–13, 16–19,
    22, 25, 28, 34, 57, 67, 76,
    86–8, 95–6, 115, 119, 156,
    158–61, 163–4, 181, 186, 193,
    231, 235–6
Russell's paradox (or contradiction)
    12–13
Ryle, G. 22

Saccheri, G. 79–80, 83–4
Schilpp, P. A. 178
Schlick, M. xii, 17–21, 24, 156,
    163, 175–7
Schrödinger, E. 20, 61, 199
Schumacher, H. K. 82
scientific conception of the world
    v, 17, 19–20, 24
scientific method vii, 3, 6, 29,
    32–5, 55, 205, 231, 234
scientific revolution(s) 38, 65, 68,
    153
severe testing, principle of 215
Sheffer stroke symbol 171
Sidgwick, H. 231
Simon, H. A. 32, 69, 71, 234
simple observation statement(s)
    167–73
simplicity 211, 214, 237
Skidelsky, R. 231
Skolem, T. 13
Snell, W. 110
Sraffa, P. 186
Stefan, J. 223
sulphonamide drugs vii, xii, 36,
    48–9, 51, 234
synthetic 76–7, 86, 88, 95,
    108–10, 235

tacking paradox 210, 217
telescope 7, 134
Thagard, P. 237

theory-laden viii, 132, 137–44,
    146–7, 236
Thomson, W. 195
Tolstoy, L. 161, 184
Toulmin, S. 168
Tractatus viii, 19, 142, 155–6, 160,
    163–8, 171–2, 174–5, 177,
    181–3, 185–7, 203

uniformity of nature vii, 8–9, 28,
    34
Uranus 100, 111
Urbach, P. 209

verifiability criterion 169, 171–2,
    177–80, 210–11, 213
Vienna Circle v, vii, viii, xii, 3, 8,
    10–11, 13, 17–25, 57, 64, 67,
    76, 109, 119, 132, 153, 155–6,
    161, 163–5, 167–9, 171–5,
    177–8, 180, 187, 189, 191, 193,
    232
Vienna Secession 157

Voltaire 67
Von Wright, G. H. 186
Vuillemin, J. 110, 193, 235

Waismann, F. 17, 19, 163, 175–7
Waites, B. 155
Walter, B. 157
Weber, B. xiv
Whitehead, A. N. 12, 14, 16, 19,
    57
Whittaker, E. 232
Wien, W. 223–4
Wittgenstein, K. 157
Wittgenstein, L. viii, 18–19,
    21–3, 122, 124, 142, 153,
    155–68, 171–2, 174–7, 181–8,
    193, 203, 228, 236
Wright, Sir Almroth 43–4

Zahar, E. 232
Zermelo, E. 13
Zytkow, J. M. 69, 71